Chemistry of Solid State Materials

Chemical synthesis of advanced ceramic materials

Chemistry of Solid State Materials

Series Editors
A. R. West, Reader in Chemistry, University of Aberdeen
H. Baxter, formerly at the Laboratory of the Government Chemist,
London

Chemical synthesis of advanced ceramic materials

David Segal

Materials Chemistry Department, Harwell Laboratory, Oxfordshire

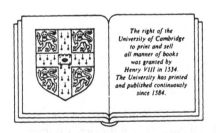

The right of the
University of Cambridge
to print and sell
all manner of books
was granted by
Henry VIII in 1534.
The University has printed
and published continuously
since 1584.

Cambridge University Press

Cambridge

New York   Port Chester   Melbourne   Sydney

CAMBRIDGE UNIVERSITY PRESS
Cambridge, New York, Melbourne, Madrid, Cape Town, Singapore,
São Paulo, Delhi, Dubai, Tokyo, Mexico City

Cambridge University Press
The Edinburgh Building, Cambridge CB2 8RU, UK

Published in the United States of America by
Cambridge University Press, New York

www.cambridge.org
Information on this title: www.cambridge.org/9780521424189

First published 1989
First paperback edition 1991

A catalogue record for this publication is available from the British Library

Library of Congress Cataloguing in Publication Data

Segal, David, 1950–
Chemical synthesis of advanced ceramic material / David Segal.
      p.   cm.
Bibliography: p.
Includes index.
ISBN 0 521 35436 6
1. Ceramics.   I. Title.
TP815.S465 1989
666—dc 19   88–27537 CIP

ISBN 978-0-521-35436-3 Hardback
ISBN 978-0-521-42418-9 Paperback

Dedicated to my mother and father

Contents

Contents

Contents

Preface

Advanced ceramic materials have attracted increasing attention through-out the 1980s from many disciplines including chemistry, physics, metallurgy and materials science and this multidisciplinary approach is illustrated by the diverse range of journals and conferences where information is disseminated. In addition the discovery of high-temperature ceramic superconductors in 1986 has raised the profile of advanced ceramics activities not only within the scientific community but also among the general public. Attendance at conferences and surveys of scientific literature show that chemical synthetic methods have played an increasing role, over the past fifteen years, in improving the properties of ceramic materials. Books concerned with fabrication and physical properties of ceramics do not, in my opinion, highlight chemical aspects of ceramic preparations which are not the principal interest of physical, organic and inorganic chemistry textbooks.

My discussions with undergraduate and postgraduate students in chemistry and materials science as well as university lecturers and those in industry concerned with research into and manufacture of advanced ceramics produced two conclusions. Firstly, there did not seem to be a short volume available which acted as a bridge between pure chemistry and conventional ceramic studies such as fabrication. Also, although scientific publications and conference proceedings proliferate it was not obvious how a comprehensive view of the rapid inroads chemistry is making into ceramic synthesis could be obtained. I see this book as that bridge between pure chemical and conventional ceramic studies. I have included a chapter on fabrication for continuity but this is not the main theme. I have not discussed the mechanisms and structures of all reactions and materials described here, or listed 'recipes' for ceramic synthesis. What I have attempted to show is the role chemistry has in the synthesis of advanced ceramic materials but, at all times, synthetic routes are related to the desired ceramic properties for materials in the form of powders, fibre, coatings or monoliths made on the laboratory and industrial scale. All branches of chemistry contribute to advanced ceramic development but three areas occur repeatedly throughout this book, namely colloid chemistry, homogeneous nucleation processes and chemistry at the organic–inorganic interface.

Finally, a paragraph on acknowledgements. I thank authors and copyright owners in Europe, Japan, Australia and The United States of America for giving permission to reproduce their photographs in this book. I am indebted to staff of the Harwell Library who obtained numerous scientific publications for me while line diagrams were drawn in the Tracing Office at Harwell. Anita Harvey typed the manuscript; and lastly an acknowledgement to my employer, The United Kingdom Atomic Energy Authority for permission to publish this book.

Harwell Laboratory   David Segal

Symbols

N	Avogadro number
h	Planck's constant
k	Boltzmann constant
e	Electronic charge
c	Ionic strength
z	Ion valency
%	Percentage
T	Absolute temperature
T_c	Superconducting transition or critical temperature
E	Young's modulus
S	Tensile fracture strength
$2C$	Crack length
C_1	BET constant
S_A	Surface area
D	Diffusion coefficient
D_c	Crystallite dimension
K	Equilibrium constant
K_{IC}	Fracture toughness
K^*	Constant in Scherrer equation (A.3)
G	Gravitational constant
η	Liquid viscosity
κ	Reciprocal double layer thickness
Q	Scattering vector
Ω	Angular rotational velocity
R, R'	Alkyl chain
R^*	Reflectivity at near normal incidence
R_G	Gas constant
λ	Wavelength
L	Nucleation rate per unit volume
I, I_0	Scattered intensity
$P(\phi)$	Shape factor
ϕ	Angle between incident and scattered radiation
V_L	London energy between two atoms
V_A	van der Waals – London energy for macroscopic bodies
V_R	Electrostatic energy of repulsion
V_S	Free energy due to adsorbed layer overlap
V_T	Total potential energy between particles
θ_B	Bragg angle

Symbols

$\theta_{1/2}$	Pure diffraction broadening at half-peak height
θ	Contact angle at solid-air-liquid interface
I_R	Rayleigh scattered intensity
I_{RG}	Rayleigh–Gans scattered intensity
ψ_0	Surface charge
ν_0	Ground-state electron vibrational frequency
α	Static atomic polarizability
λ^*	$= 3h\nu_0\alpha^2/4$
$A, A_1,$	
A_2, A_{12}	Hamaker constant
a	Sphere radius
H_0	Separation of spheres or flat plates at closest approach
x	$= H_0/2a$
q	Number of molecules per unit volume of material
ε_0	Permittivity of vacuum
ε	Relative permittivity of medium
l	Side length of a cube
l_0	Sample to detector distance
l_1	Sedimentation distance
l_2	Distance between particle and axis of rotation
l_3	Radius of rotation for particles
$n_0, n_1, n_2,$	
n_p, n_m	Refractive index
n	Number of molecules in critical cluster
$\Delta G'_n$	Free energy of formation for critical cluster
r	Distance between atoms
r_p	Pore radius
r_d	Radius of liquid droplet
r_g	Radius of gyration
r_n	Critical radius of cluster
b, f, g, j, δ	Number of moles
$t_{1/2}$	Half-life
$t, t_1, t_2,$	
t_3, t_c, τ	Time
ρ, ρ_p, ρ_1	Density
γ	Surface tension of liquid
γ'	Surface energy
Δ_p	Capillary pressure
$p_F, p_G,$	
p_j	Partial gas pressures
p^0	Saturation vapour pressure
p'	Vapour pressure
p	Gas pressure
M_w	Molecular weight
$\bar{M}_w$	Weight average molecular weight
$\bar{M}_n$	Number average molecular weight
h_c	Coating thickness
C_x, C_0	Concentration of solute or reactant
C_s	Saturated solute concentration

Symbols

C_{ss}	Supersaturated solute concentration
N_p	Number of particles
β	Overall growth coefficient for droplets
ν	Molar volume of liquid phase
ν_l	Volume of molecule in liquid phase
ν_a	Adsorbate volume at a specified relative pressure
ν_m	Adsorbate volume for monolayer coverage per unit mass of solid
m	Mass of a molecule
m_d	Volume of a diffusing vacancy
M	Metal

1 Introduction: the variety of ceramic systems

1.1 Introduction

The international advanced ceramics industry is concerned with basic research and ceramic fabrication as well as manufacture of powders and fibres while the success of research can be measured by its application to large-scale economic production of ceramics which function in particular working environments. Advanced ceramic materials are defined in this introductory chapter and their variety and uses are explained. The ceramics industry is large and an indication of its volume production and monetary value is also given here. Finally, the recent discovery of high-temperature oxide superconductors has had a tremendous impact on worldwide ceramic activities and a section is included on the properties and potential applications of these advanced ceramic materials.

1.2 From traditional to advanced ceramics

Ceramics are the group of non-metallic inorganic solids and their use by man dates from the time of ancient civilisations. In fact, the word ceramic is of Greek origin and its translation (keramos) means potter's earth. Traditional ceramics are those derived from naturally occurring raw materials and include clay-based products such as tableware and sanitaryware as well as structural claywares like bricks and pipes. Also in this category are cements, glasses and refractories. Examples of the latter are chrome–magnesite refractories used in the steel-making industry and derived from magnesite ($MgCO_3$) and chrome ore. Advanced ceramics are developed from chemical synthetic routes or from naturally occurring materials that have been highly refined. A variety of names has been used to describe ceramic systems. Hence, advanced ceramics are also called engineering ceramics whereas the phrases 'special', 'fine' and 'technical' have all been used in connection

1

with these materials. When their use depends on mechanical behaviour, advanced ceramics are sometimes referred to as structural components whereas electroceramics are a class of advanced ceramic whose application relies on electrical and magnetic properties. Books by Norton (1968) and Shaw (1972) contain further details on traditional ceramics whereas Morrell (1985) has described the classification of ceramic systems.

1.3   Structural and refractory applications of engineering ceramics

Structural components derived from engineering ceramics are used as monoliths, coatings and composites in conjunction with or as replacements for metals when applications rely on mechanical behaviour of the ceramics and their refractory properties, that is chemical resistance to the working environment. Physical properties of ceramics and metals are compared in table 1.1 some of their magnitudes are shown in tables 1.2 and 1.3. Nickel superalloys are currently the main high-temperature materials for components such as combustors in gas turbine engines (Meetham, 1986). They have melting points around 1573 K and a maximum working temperature near 1300 K. Cast iron parts in reciprocating (i.e. petrol and diesel) engines have properties shown in table 1.3. Compared with metals, ceramics are generally more resistant to oxidation, corrosion, creep and wear in addition to being better thermal insulators. They have higher melting points (table 1.4) and greater strength than superalloys at elevated temperature so that a major potential application, particularly for silicon nitride, is in gas turbine and reciprocating engines where operating temperatures higher than attainable with metals can result in greater efficiencies. This enhanced strength is shown in figure 1.1 for hot-pressed silicon nitride (HPSN), hot-pressed silicon carbide (HPSC), hot isostatically pressed silicon nitride (HIPSN), sintered silicon nitride (SSN), sintered silicon carbide (SSC), reaction-bonded silicon nitride (RBSN) and reaction-bonded silicon carbide (RBSC). Although ceramics offer improvements in engine efficiency, incorporation of silicon nitride over the past three decades has been slow, mainly because of the difficulty in reproducible fabrication of dense components to close dimensional tolerances.

Silicon nitride occurs in two phases, the α and the β forms. The β form, whose structure is shown in figure 1.2, consists of SiN_4 tetrahedra joined together by sharing corners in a three-dimensional network. It is

Table 1.1. *Relative properties of ceramics and metals (AE Development, 1985)*

Property	Ceramics	Metals	Ratio, property of ceramics: property of metal
Ductility	Very low	High	$(0.001-0.01):1$
Density	Low	High	$0.5:1$
Fracture toughness	Low	High	$(0.01-0.1):1$
Young's modulus	High	Low	$(1-3):1$
Hardness	High	Low	$(3-10):1$
Thermal expansion	Low	High	$(0.1-0.3):1$
Thermal conductivity	Low	High	$(0.05-0.2):1$
Electrical resistance	High	Low	$(10^6-10^{10}):1$

Table 1.2. *Physical properties for alloys, oxide and non-oxide ceramics (Briscoe, 1986)*

Material	Specific gravity (kg m^{-3})	Thermal expansion coefficient $(10^{-6}\,\text{K}^{-1})$	Thermal conductivity $(\text{W m}^{-1}\,\text{K}^{-1})$
Alumina	4000	9	20
Toughened zirconia polycrystals	5800	10	2
Sintered silicon carbide	3100	4.5	40
Sintered silicon nitride	3100	3.2	12
Hot-pressed silicon nitride	3100	3.1	30
Window glass	2200	9	1
Aluminium alloys	2800	22	146
Nimonic superalloys	8500	15	16

possible to replace silicon by aluminium and maintain charge neutrality in the crystal lattice by substitution of nitrogen with oxygen. The resulting solid solutions in the Si–Al–O–N system are known as β'-sialons (K. H. Jack, 1986) whose structures are identical with β-Si$_3$N$_4$ over the composition range $Si_{6-b}Al_bO_bN_{8-b}$ $(0<b<4)$. They exhibit mechanical behaviour similar to β-Si$_3$N$_4$ and have some features of aluminium oxide. However, in contrast with Al$_2$O$_3$, which consists of six-coordinated Al,

Table 1.3. *Mechanical properties of cast irons, oxide and non-oxide ceramics (Lackey et al., 1987)*

Material	Young's modulus GPa	Fracture toughness (MPa m$^{1/2}$)	Fracture strength at room temperature MPa
Alumina	380	2.7–4.2	276–1034
Partially stabilised zirconia	205	8–9 at 293 K	
		6–6.5 at 723 K	600–700
		5 at 1073 K	
Sintered silicon carbide	207–483	4.8 at 300 K	96–520
		2.6–5.0 at 1273 K	
Sintered silicon nitride	304	5.3	414–650
Hot-pressed silicon nitride	304	4.1–6.0	700–1000
Glass ceramics	83–138	2.4	70–350
Pyrex glass	70	0.75	69
Cast irons	83–211	37–45	90–1186

Table 1.4. *Melting point and maximum working temperatures of ceramics (Lay, 1983)*

Material	Melting point or decomposition temperature/K	Maximum working temperature/K	
		In oxidising atmosphere	In reducing atmosphere
Alumina	2323	2173	2173
Stabilised zirconia	2823	2473	
Silicon carbide	2873	1923	2593
Boron carbide	2723	873	2273
Tungsten carbide	3023	823	2273
Reaction-bonded silicon nitride	2173	1473	2143
Boron nitride	2573	1473	2473
Titanium diboride	3253	1073	>2273

β'-sialon contains Al that is four-coordinated by oxygen and this results in an enhanced Al–O bond strength compared with the oxide. Unlike Si_3N_4, β'-sialons can be densified readily by pressureless sintering and they have been put into commercial production by Lucas Cookson

Structural and refractory applications

Syalon Limited. Syalon components shown in figure 1.3 include automotive parts such as valves, valve guides and seats, tappets, rocker inserts and precombustion chambers in addition to weld shrouds, location pins, extrusion dies, tube drawing dies and plugs.

Aluminium titanate is used as port liners in some automobile engines because its low thermal conductivity (2 W m^{-1} K^{-1}) reduces heat flow to the cylinder block and hence the amount of cooling required. Glass ceramics have applications (table 2.1) in cooking utensils, tableware, heat exchangers, vacuum tube components and missile radomes. Partially stabilised zirconia was developed in 1975 at the Australian Commonwealth Scientific & Industrial Research Organisation (CSIRO) and is nowadays manufactured by Nilcra-PSZ Limited. This material is particularly suited for withstanding mechanical and thermal shock because of its high fracture toughness (table 1.3). Examples are dies for

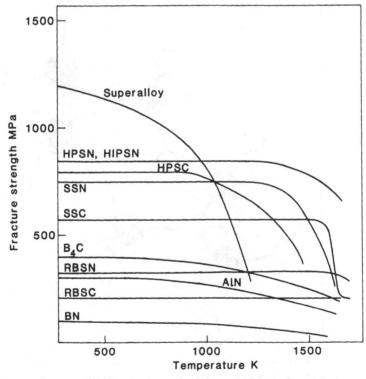

Figure 1.1. Variation of strength with temperature for non-oxide ceramics (Heinrich, 1985).

5

extrusion of copper and aluminium tubes, diesel engine cam follower faces, valve guides, cylinder liners and piston caps, wear and corrosion-resistant nozzles in papermaking equipment, wear resistant inserts such as tabletting dies as well as scissors and knives.

Not all ceramic components require high-temperature strength. The high Young's modulus (550 GPa) of titanium diboride, TiB_2, makes it useful for armour plating (Knoch, 1987) whereas ceramics are suitable materials in seals because of their chemical resistance (table 1.5). Hence sintered silicon carbide is used for mechanical seals and sliding bearings whereas boron nitride, which is not wetted by glass and liquid metals, constitutes break rings in the horizontal continuous casting process for steels. Boron carbide, a harder ceramic than SiC, is suited to wear-resistant applications such as grit blasting nozzles whereas Si_3N_4 is also

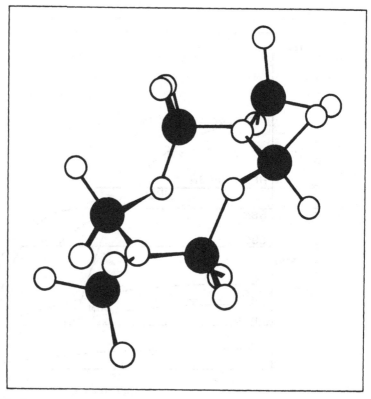

Figure 1.2. Crystal structure of β-Si_3N_4 and β'-$(Si,Al)_3(O,N)_4$. ●, metal atom, ○, non-metal atom (K. H. Jack, 1986).

used as ball bearings. An established industrial use for engineering ceramics is as cutting tools for steels where high-temperature hardness of sialons and zirconia toughened alumina together with their low reactivity towards metals are desirable properties; this application has recently been reviewed by D. H. Jack (1986).

Worldwide markets for ceramic coatings in 1985 have been estimated at around 1100 US$M (Charles H. Kline, 1987) and their breakdown is shown in figure 1.4. About 45% of this market is in optical coatings,

Table 1.5. *Resistance of ceramics towards acids and bases (Lay, 1983)*

Material	Resistance to acids	Resistance to bases
MgO, ThO_2	Lowest	Highest
BeO		
Al_2O_3, Cr_2O_3, ZrO_2	↓	↓
SiO_2, TiO_2, SiC, B_4C, Si_3N_4	Highest	Lowest

Figure 1.3. Examples of Syalon components. (Courtesy of Lucas Cookson Syalon Ltd.)

the remainder for thermal, wear and corrosion resistance. For example the thermal barrier coating in figure 1.5 contains an outer ZrO_2 layer on the combustion chamber from the RB211 gas turbine engine (Meetham, 1986) and produced an increase in combustor lifetime because of a reduction in substrate temperature by 50 K.

Bioceramics are a class of advanced ceramics that require high strength. Thus, alumina for artificial hip joints, hydroxyapatite for surgical implants, calcium phosphate as an aid to rejuvenation of bone and carbon surgical implants are used in this area which has been reviewed by Boretos (1987).

Physical data for traditional and advanced ceramics are dispersed throughout the technical literature but a useful source is available (American Ceramic Society, 1987) which collates this information for some materials described in later chapters of this book.

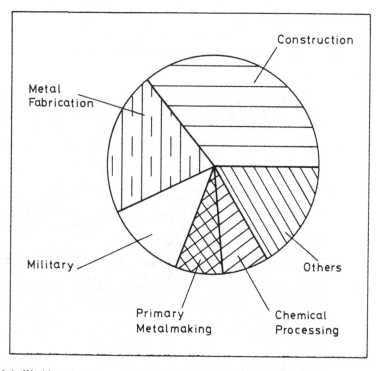

Figure 1.4. World market for ceramic coatings by end use industries, 1985 (Charles H. Kline and Company, 1987).

Figure 1.5. Thermal barrier-coated RB211 combustion chamber. (Courtesy of Rolls Royce plc.)

1.4   Electroceramics

Electroceramics or electronic ceramics can be considered as materials whose uses rely primarily on their electrical or magnetic properties rather than mechanical behaviour although optical properties are important for opto-electronic devices. The electroceramics industry is a mature one and dates from the 1940s when insulating properties of Al_2O_3 were utilised in spark plugs but nowadays this is a rapidly expanding area of materials science and forms the largest sales sector (in monetary value) for advanced ceramics. Examples of electroceramics include zinc oxide for varistors, lead zirconium titanate (PZT) for piezoelectrics, barium titanate in capacitors, tin oxide as gas sensors, lead lanthanum zirconium titanate (PLZT) and lithium niobate for electro-optic devices; physical properties required for these applications are shown in table 1.6. In the case of packaging, ceramic substrates such as Al_2O_3 are used for supporting electronic chips and interconnections. The trend towards greater miniaturisation and densification of components demands substrates with high thermal conductivity and the magnitude of this property for packaging materials is listed in table 1.7.

European and North American consumption of electronic ceramics for 1985 are compared in table 1.8, while Bell (1987) has quoted Japanese sales of electroceramics for 1983 totalling around 1500 US$M. This industry is a large volume producer as shown in table 1.9 for a range of ceramic sensors whose operation is described in more detail by Kulwicki (1984). At the present time electroceramics such as piezoelectric knock sensors and thermistors for fuel gauges are being increasingly used in automobiles (Taguchi, 1987), while Tuttle (1987) has reviewed their application in opto-electronic devices. High-temperature oxide superconductors, a class of electronic ceramics that has been recently discovered, are described below.

1.5   High-temperature oxide superconductors

Superconductors are materials that have no electrical resistance. The phenomenon was discovered by Onnes (1911a, b) who observed a superconducting transition or critical temperature, T_c, of 4.2 K for mercury in liquid helium. A gradual increase in T_c values took place over the following 75 years and the highest critical temperature up to 1986, 22.3 K, was for films of an almost stoichiometric niobium–germanium

Table 1.6. *Properties and applications of electronic ceramics (Bell, 1987)*

Material	Properties	Applications
Al_2O_3, AlN, BeO	Low permittivity; high thermal conductivity	Packaging, substrates
$BaTiO_3$	High permittivity; high breakdown voltage	Capacitors
PZT, $BaTiO_3$, $LiNbO_3$	High piezoelectric coefficients	Piezoelectric transducers, saw devices
$BaTiO_3$ (PTC)	Change of resistance with temperature	Thermistors
ZnO	Change of resistivity with applied field	Varistors
PLZT, $BaTiO_3$, $LiNbO_3$	Change of birefringence with field	Electro-optics
ZrO_2	Ionic conductivity	Gas sensors
SnO_2	Surface-controlled conductivity	Gas sensors
Ferrites	Permeability, coercive field	Magnets
PZT	Change of polarisation with temperature	Pyroelectrics

Table 1.7. *Properties of ceramics for electronic packaging (Tummala & Shaw, 1987)*

Material	Permittivity	Thermal expansion coefficient $(10^{-6}\,K^{-1})$	Thermal conductivity $(W\,m^{-1}\,K^{-1})$
Al_2O_3	9.0–9.5	6.7	25–35
Si	—	2.7	150–160
AlN	8.5–10.0	3.3	140–170
BeO	6–7	9	220–240
Si_3N_4	6.0–10.0	2.3	25–35
SiC	20–40	3.5	120–250

Table 1.8. *United States and European consumption of electronic ceramics in 1985 (Cantagrel, 1986)*

Application	North America US $M	Europe US $M	Selling price per kg of ceramic/US $
Disc capacitor	70	50	—
Multilayer capacitor	600	150	1100
Soft ferrites	90	80	3.3
Positive temperature coefficient thermistors (PTC)	30	18	330
Negative temperature coefficient thermistors (NTC)	20	6	330
Zinc oxide varistors	22	18	33

alloy, Nb_3Ge (Gavaler, 1973); other alloys also exhibited superconductivity. Three superconducting oxide systems were discovered but had T_c values below that for the Nb_3Ge alloy. Thus, substituted and reduced strontium titanates of general formula $Ba_bSr_{1-b}TiO_3$ and $Ca_jSr_{1-j}TiO_3$ had transition temperatures less than 1 K when $b \leq 0.1$ and $j \leq 0.3$ (Frederikse *et al.*, 1966). Lithium titanates, $Li_{1+b}Ti_{2-b}O_4$ have a spinel structure when $-0.2 < b < 0.33$ (Johnston *et al.*, 1973) and T_c values ranging from 7 K to 13.7 K whereas the perovskite structure for polycrystalline

Table 1.9. *Volume production of ceramic sensors (Kulwicki, 1984)*

Sensor	Number per year (millions)
Piezoelectric devices	300
PTC thermistors	200
NTC thermistors	50
ZnO varistors	100
Oxygen sensors	2
Humidity sensors	< 1
Combustible gas sensors	< 1

$BaPb_{1-b}Bi_bO_3$ had critical temperatures up to 13 K when $0.05 < b < 0.3$ but were semiconductors when $0.35 < b < 1.0$ (Sleight, Gillson & Bierstedt, 1975).

Before 1986, books and publications concerned with advanced ceramics contain few references to oxide superconductors. However, in that year Bednorz & Müller (1986) tentatively claimed that polycrystalline $La_{5-b}Ba_bCu_5O_{5(3-j)}$ ($b = 0.75, 1.0, j > 0$) exhibited a transition temperature of 30 K and high-temperature superconductivity was confirmed by . magnetic susceptibility measurements (Bednorz, Takashige & Müller, 1987). One of three phases in this ceramic was responsible for superconductivity and was identified (Takagi *et al.*, 1987) as having a potassium nickel fluoride (K_2NiF_4) perovskite structure with composition $(La_{1-b}Ba_b)_2CuO_{4-\delta}$ where $\delta < 1$ and $0.05 < b < 0.15$; La_2CuO_4, corresponding to $b, \delta = 0$, was not a superconductor. Soon after this discovery Chu and co-workers (Wu *et al.*, 1987) made a polycrystalline oxide with T_c between 80 and 93 K and which was superconducting at 77 K in liquid nitrogen. The superconducting phase was shown to be an yttrium barium cuprate, $YBa_2Cu_3O_{7-\delta}$ ($0 < \delta < 1$), later termed the 1–2–3 phase, whose structure, an orthorhombic oxygen deficient perovskite is shown in figure 1.6. This structure has two important features: oxygen vacancies in the lattice and the presence of copper in the Cu^{2+} and less-common Cu^{3+} valence states. While high-T_c oxide superconductors are a novel class of ceramic material, oxygen-deficient perovskites are not new (Raveau & Michel, 1987; Michel & Raveau, 1984). Reversible intercalation of oxygen into the latter makes them suitable as sensors in applications when the electronic properties of materials are sensitive to their oxygen content.

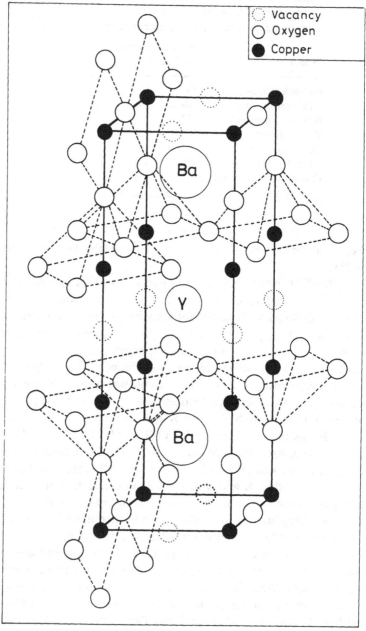

Figure 1.6. Orthorhombic crystal structure of the yttrium barium cuprate high-T_c ceramic superconductor (Clarke, 1987).

High-temperature oxide superconductors

Ceramics research does not often occupy the public and scientific limelight but discovery of high-temperature oxide superconductors caused unprecedented worldwide research interest throughout 1987. International conferences attracted thousands of participants, the ceramics were frequently described in national newspapers, scientific articles were published at the rate of around 100 a month and national initiatives on these materials are being set up in several countries. For example, in the United Kingdom a National Superconductivity Centre has been established at Cambridge University. The intense, almost frenzied interest arose from potential applications of superconducting ceramics and the temperature at which they develop superconductivity.

Because superconductors have zero electrical resistance, no energy losses occur when an electric current is transmitted through them. Superconductors also repel all magnetic fields; this repulsion, known as the Meissner effect, is illustrated in figure 1.7, which shows a samarium–cobalt magnet levitated by an yttrium barium cuprate pellet (diameter 1.0 cm) immersed in liquid nitrogen. Superconducting metal alloys are used for production of magnetic fields up to *ca.* 15 tesla in applications of

Figure 1.7. Meissner effect for the yttrium barium cuprate high-T_c ceramic superconductor. (Courtesy of the UKAEA.)

nuclear magnetic resonance, including chemical analysis and the tomographic scanning technique in medicine, but the high cost of liquid helium coolant restricts their wider use. However, liquid nitrogen is approximately 30 times cheaper than liquid helium and easier to handle. Commercial exploitation of ceramic superconductors will be easier as T_c values increase. Potential applications include magnetically levitated trains, power transmission without energy losses, magnetic shielding, very fast computers, compact powerful electric motors with high efficiency and magnets for nuclear fusion.

Successful exploitation of ceramic superconductors will be achieved only if formidable problems are overcome. Current densities around 10^5 A cm^{-2} are required, 100 times greater than has been measured so far and they must be able to withstand magnetic fields as high as 20 tesla without losing their superconductivity. In addition, oxide superconductors will be used in particular shapes such as wires, coils (Jin *et al.*, 1987) and coatings (Wen *et al.*, 1987) but they are brittle solids and need to be fabricated with high fracture strength and toughness. The ceramics also absorb water vapour from the atmosphere, a process that has a detrimental effect on their superconducting properties. However, if these difficulties are surmounted then worldwide markets for high-temperature oxide superconductors, which may have a profound effect on everyday life, could be very large.

2 Conventional routes to ceramics

2.1 Introduction

Precipitation from solution, powder mixing and fusion are all conventional techniques for synthesis of traditional and advanced ceramics on both the laboratory and industrial scale. These methods, with selected examples, are described in this chapter together with limitations of their use for advanced ceramic materials. There is increasing demand for alternative routes to ceramic materials that impart superior properties compared with those attainable from conventional syntheses and the reasons for this continuing search for other synthetic pathways are also described here.

2.2 Precipitation from solution

Alumina occurs as the mineral bauxite and is refined in the Bayer process whereby ore is initially dissolved under pressure in sodium hydroxide so that solid impurities (SiO_2, TiO_2, Fe_2O_3) separate from sodium aluminate solution (Evans & Brown, 1981). This solution is either seeded with gibbsite crystals (α-Al_2O_3.$3H_2O$) or undergoes auto-precipitation to bayerite (β-Al_2O_3.$3H_2O$) after its neutralisation with CO_2 gas. Temperature, alumina supersaturation and amount of seed affect particle size during crystallisation but, as for other precipitation reactions, the product is agglomerated (section 9.3). Alumina production from the Bayer process was 31.9×10^6 Mg in 1985; 90 weight % of this was reduced to metal and only 5 weight % found ceramic and refractory applications (MacZura, Carbone & Hart, 1987).

World output of refined zirconia, which occurs as the minerals baddeleyite (ZrO_2) and zircon sand (ZrO_2.SiO_2), is around 2×10^4 Mg per annum (Clough, 1985). Zirconyl chloride is obtained from the product of fusion between zircon and sodium hydroxide and a manufacturing

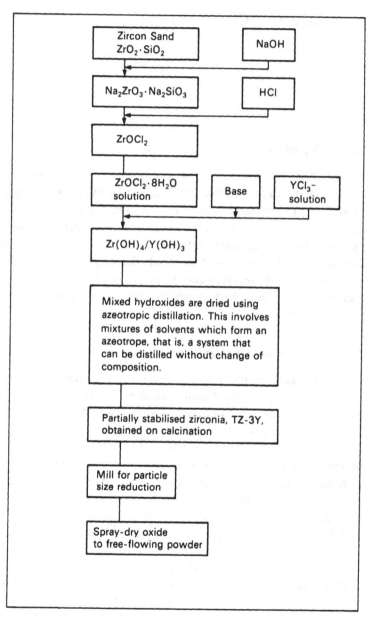

Figure 2.1. Schematic representation of a commercial process for producing partially stabilised zirconia powder by a coprecipitation process (Toya Soda Company, 1984).

route for partially stabilised zirconia powder (section 3.9) involving precipitation from this chloride solution is shown in figure 2.1. Problems can
arise when two or more components are coprecipitated. Thus, different
species do not always deposit from solution at the reaction pH, while
washing procedures can selectively remove a precipitated component as
well as dissolve entrained electrolyte. The difficulty in maintaining
chemical homogeneity is serious as inhomogeneities have a deleterious
effect on the mechanical and electrical properties of ceramics. Because
precipitation results in agglomerated powders, grinding, dry-milling or
wet-milling with water or a non-aqueous liquid are used for particle size
reduction so that powder compacts will sinter to near theoretical density.
These comminution processes can introduce impurities into the ceramic
from the grinding media (e.g. alumina, zirconia or stainless steel balls)
while high temperatures are required for densification, e.g. 1873 K for
calcium titanate ($CaTiO_3$) compacts derived from coprecipitated
calcium peroxide and titanium dioxide (Yang, Yamada & Miller, 1985).

Precipitation reactions are not restricted to oxides and hydroxides.
Hence, for the high-T_c oxide superconductor $La_{1.85}Ba_{0.15}CuO_4$
(Jorgensen *et al.*, 1987), La, Ba and Cu oxalates were deposited from
electrolyte solutions and sintered in air at 1373 K. Because these
materials reversibly intercalate O_2, the annealing temperature and rate
of cooling, which affect their superconducting properties and the $Cu^{3+}/
Cu^{2+}$ ratio (section 1.5) must be carefully controlled.

2.3   Powder mixing techniques

Multicomponent oxide powders are synthesised from conventional
mixing techniques by initially blending together starting materials,
usually metal oxides and carbonates, after which the mixtures are
ground or milled. Comminuted powders are then calcined, sometimes
after compaction, and the firing sequence may be repeated several times
with intermediate grinding stages. As for coprecipitation, impurities can
be introduced into the ceramic from the grinding operation; grinding
also results in angular-shaped powders.

Positive temperature coefficient thermistors have been produced by
these techniques (Philips Electrical Industries, 1954). Hence for the composition 32.13 mole % BaO; 17.32 mole % CaO; 50.5 mole % $TiO_2 + 0.5$
atomic % La individual components were ball-milled together, heated
initially at 1273 K and then sintered at 1673 K after shaping by an extrusion method.

Conventional routes to ceramics

Several problems are associated with mixing powders. High temperatures required for reaction between components can result in loss of volatile oxides, while milling may not comminute powders sufficiently for complete reaction to occur on calcination. It is difficult to obtain reproducible uniform distributions of material in ball-milled powders especially when one fraction is present in small amounts as occurs in electroceramics whose properties are often controlled by grain boundary phases containing minor quantities of additives. The yttrium barium cuprate superconductor $YBa_2Cu_3O_{7-\delta}$ was synthesised (Cava et al., 1987) by mixing Y_2O_3, $BaCO_3$ and CuO, grinding and heating at 1223 K in air. Powder was then pressed into pellets, sintered in flowing O_2, cooled to 473 K in O_2 and removed from the furnace. For non-oxide ceramics, conventional mixing techniques involve preparation of dry feeds like SiO_2–C mixtures for carbothermic reduction (section 9.4).

In the manufacture of fuel for thermal nuclear reactors (Wilkinson, 1981) uranium hexafluoride, UF_6, is converted to UO_2 powder

$$UF_6 + H_2 + 2H_2O \xrightarrow{873\ K} UO_2 + 6HF \qquad (2.1)$$

after which the powder is granulated and sintered to high-density UO_2 pellets. Powder mixing is a dusty process and can create radiological hazards for certain nuclear fuels although dust-free methods for preparation of fuels are described in section 4.6.

2.4 Fusion routes to ceramics

Fusion techniques are used for traditional ceramics such as window glass and advanced ceramics such as abrasive grain (figure 4.12). These methods are limited by the melting point of reactants (oxides and carbonates) while high temperatures can lead to loss of volatile oxides, for example PbO from the melt (Cable, 1984). Glass ceramics were discovered by Stookey (1964) and are polycrystalline materials made by controlled crystallisation of glasses. Examples of these materials are shown in table 2.1. Reactants are mixed as powders and melted, after which nucleating agents are added. These promote nucleation at temperatures with corresponding viscosities between 10^{10}–10^{11} Pa s, which are usually about 50 K above the softening temperature of the glass (Doyle, 1979). The temperature is then raised until crystallisation occurs and the microstructure is developed.

20

Table 2.1. *Nucleating agents and applications for glass ceramics (Doyle, 1979)*

System	Nucleating agents	Crystal phases	Applications
$Li_2O–Al_2O_3–SiO_2$ $Al_2O_3 > 10\%$	TiO_2 $TiO_2 + P_2O_5 + ZrO_2$	β-spodumene β-eucryptite	Cooking utensils; heat exchangers; telescopic mirrors
$Li_2O–Al_2O_3–SiO_2$ $Al_2O_3 < 10\%$	P_2O_5 Cu, Ag, Au	Lithium disilicate, quartz, lithium metasilicate	Glass ceramic/metal seals Photochemical machining
$Li_2O–ZnO–P_2O_5–SiO_2$	P_2O_5	Lithium disilicate, lithium zinc silicate, quartz	Vacuum tube components
$Na_2O–BaO–Al_2O_3–SiO_2$	TiO_2	Nepheline, hexacelsian	Tableware
$MgO–Al_2O_3–SiO_2$	TiO_2	Cordierite, cristobalite	Missile radomes

2.5   The need for improved synthetic routes to advanced ceramics

Advanced ceramics are usually manufactured by conventional processes and it should not be construed in any way that these are substandard materials. However, there is increasing interest in alternative synthetic routes for several reasons. First, chemical impurities have a deleterious effect on high-temperature mechanical behaviour of engineering ceramics and electrical properties of electroceramics so that greater purity derived from novel syntheses can lead to improved physical properties. Secondly, conventional routes are not versatile for producing a wide range of coatings and fibrous materials. Non-uniform powder compositions make reproducible component fabrication difficult because of chemical inhomogeneity and voids in the microstructure. Hence a requirement arises for powders with physical characteristics that allow reliable fabrication. Finally, there is no intrinsic reason why only powder feedstocks should be suitable for manufacturing monolithic ceramics. Therefore methods for achieving these, with other inter-mediates, are described in later chapters as are non-conventional syntheses of advanced ceramic materials.

3 Ceramic fabrication

3.1 Introduction

Solid ceramic bodies are generally produced by using the process of powder compaction followed by firing at high temperature. Sintering or densification occurs during this heat treatment and is associated with joining together of particles, volume reduction, decrease in porosity and increase in grain size. The phase distribution or microstructure within the ceramic is developed during sintering and fabrication techniques used for shaping ceramics are described in this chapter. The aim of these techniques is to produce microstructures suitable for particular applications. Hence a fine-grained distribution is required for strength. Controlled grain size is necessary where optical properties such as translucency are required. Strength and toughness of ceramic systems are also discussed here with particular reference to the role of powder preparation on the strength of ceramics.

3.2 Solid-state sintering

The driving force for sintering is reduction in surface free energy associated with a decrease of surface area in powder compacts due to removal of solid–vapour interfaces. Vapour-phase nucleation is described in chapter 9 by using the Kelvin equation (9.1), which is also applicable to mass-transport processes in a consolidated powder (Kingery, 1983). The vapour-pressure difference across a curved interface can enhance evaporation from particle surfaces and condensation at the neck between two particles, particularly for particle diameters of several micrometres or less, such as occur in ceramic fabrication. Although this evaporation–condensation process produces changes in pore shape and joins particles together, the centre-to-centre distance between particles remains constant so that shrinkage and densification

Table 3.1. *Alternative pathways for mass transport during sintering (Kingery et al., 1976)*

Transport pathway	Mass source	Mass sink
Vapour transport	Surface	Neck
Surface diffusion	Surface	Neck
Lattice diffusion	Surface	Neck
Boundary diffusion	Grain boundary	Neck
Lattice diffusion	Grain boundary	Neck
Lattice diffusion	Dislocations	Neck

do not occur. The driving force for mass transport by solid-state processes shown in table 3.1 for ceramic powders with low vapour pressure is the difference in free energy between the neck region and surface of particles. As for the evaporation–condensation pathway, transport from surface to neck by surface and lattice diffusion does not cause densification. This is produced only by diffusion from the grain boundary between particles and from the bulk lattice. Covalent ceramics such as Si_3N_4 are more difficult to sinter to high density than ionic solids, for example Al_2O_3, because of lower atomic mobilities, although difficulties can be overcome by using very fine powders *ca.* 0.1 μm diameter, high temperature and high pressures (Popper, 1983).

Impurities such as oxygen and chlorine in Si_3N_4 often migrate during sintering to grain boundaries where they reduce the interfacial surface energy and impair densification, creep behaviour, oxidation resistance and high-temperature strength; preparation of non-oxide powders with low impurity levels are described later in chapters 6 and 9.

When mass transport takes place from a grain boundary to the neck by lattice diffusion the linear shrinkage or sintering rate for a powder compact can be written as (Kingery, Bowen & Uhlmann, 1976)

$$\text{Sintering rate} = (20\gamma' m_d D/1.4kT)^{0.4} a^{-1.2} t^{0.4} \tag{3.1}$$

where γ' is the surface energy, m_d the atomic volume of a diffusing vacancy, D the self-diffusion coefficient, k the Boltzmann constant, T the absolute temperature, a the particle radius and t the time for sintering. Linear shrinkage is approximately proportional to the inverse of the particle radius but is not greatly affected by sintering time so that small particles (*ca.* 1 μm diameter) with short diffusion distances are suitable for densification. Detailed descriptions of sintering processes are given in the excellent book by Kingery *et al.* (1976). Idealised microstructures are shown in figure 3.1, while a wider range of these phase distributions has been compiled by Morrell (1985).

3.3   Uniaxial pressing

In uniaxial pressing a hard steel die is filled with either dry powder, or a powder containing up to several weight percent of H_2O, and a hard metal punch is driven into the die to form a coherent compact. Van der Waals forces (section 4.3) cause aggregation of fine powders so that binders such as polyvinyl alcohol and lubricants (e.g. stearic acid) are incorporated into them by, for example, spray-drying (section 10.5) in order to improve their flow properties and homogeneity of the product (Richerson, 1982). It is important that the unfired or green body has adequate strength for handling before the firing operation, during which

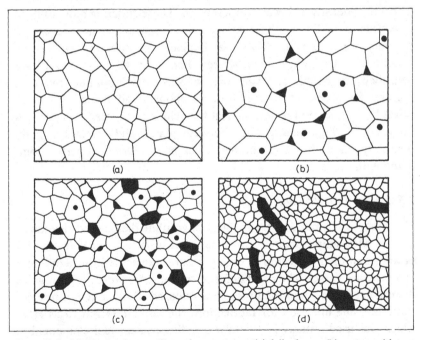

(a)

(b)

(c)

(d)

Figure 3.1. Idealised polycrystalline microstructures (a) fully dense, (b) porous with small pores, (c) porous with pores and grains of similar size and (d) porous with large pores (Davidge, 1979).

organic additives are decomposed. Uniaxial pressing can be readily automated (Thurnauer, 1958) and is particularly suited for forming components with a simple shape such as flat discs and rings that can be produced to close dimensional tolerances, thus avoiding post-firing diamond machining operations.

3.4   Hot uniaxial pressing

Hot uniaxial pressing or hot-pressing involves simultaneous application of heat and pressure during sintering. A refractory die, usually graphite, is filled with powder, which, after compaction, is heated in an inert atmosphere. Hot-pressing produces higher density and smaller grain sizes at lower temperatures compared with uniaxial pressing and is particularly suited for fabrication of flat plates, blocks and cylinders (Richerson, 1982). Stresses set up by the applied pressure on contacts between particles increase the driving force for sintering and remove the need for very fine particle sizes. Additives such as magnesium oxide and yttrium oxide, which are often used for Si_3N_4, allow achievement of theoretical density at lower temperatures. These sintering aids result in formation of a liquid phase and particle rearrangement because of capillary forces arising from the Laplace equation (5.12) and by dissolution–recrystallisation processes. Petzow & Kaysser (1984) have described, in detail, sintering mechanisms when a liquid phase is present. However, advantages brought about by additives have to be offset by degradation in mechanical behaviour of sintered components especially at high temperature because glassy and crystalline grain boundary phases derived from them often have inferior properties compared with the matrix.

3.5   Isostatic pressing

Hot-pressing is limited by the strength of graphite dies and fabrication of articles with simple shapes. In isostatic pressing, a mold is filled with powder and then subjected to high pressure transmitted through liquid in a pressure vessel. The mold deforms to compress powder and regains its original shape when the pressure is released. Both cold and hot isostatic pressing are used for fabrication, the latter referred to as HIP, which was used initially in 1955 for cladding nuclear fuel elements but currently finds many applications as shown in table 3.2. Compared with uniaxial pressing techniques, isostatic pressing yields green compacts

Table 3.2. *Uses of hot isostatic pressing (Richter, Haour & Richon, 1985)*

Category	Commercial use
Rejuvenation	Superalloys, titanium
Castings	Superalloys, titanium
Post-sintering	Ceramic tools, hard metals
Powder compaction	Superalloys, stainless steels, titanium, chromium
Joining	Fuel elements, valves, aerospace and nuclear parts
Infiltration	Carbon–carbon composites

with higher and more uniform density in a wider variety of geometries although accurate shaping is difficult. James (1983) has described isostatic pressing in more detail.

3.6   Reaction-bonding

The reaction-bonding technique is associated with Si_3N_4 and SiC ceramics. For the former, silicon articles are fabricated by a variety of processes including uniaxial pressing, injection molding and slip casting. These parts are then reacted with N_2 to form the reaction-bonded silicon nitride (RBSN) components that can be produced in complicated designs with a high degree of dimensional accuracy because no shrinkage occurs during nitridation, although the non-oxide ceramics contain about 15 volume % porosity. Because of its porosity, reaction-bonded material has inferior strength compared with hot-pressed silicon nitride. However, in a recent development Mangels & Tennenhouse (1980) showed that post-sintered reaction-bonded silicon nitride (PSRBSN) could be obtained with 95% of theoretical density and a strength and microstructure similar to hot-pressed material. In this technique, sintering aid is incorporated into RBSN, either Y_2O_3 to Si powder before nitridation or MgO by impregnating RBSN with a methanolic magnesium chloride solution after which the reaction-bonded component is heated between 2053 and 2198 K under an N_2 gas pressure of 2.1 MPa. This post-sintering technique offers near-net-shape fabrication of high-density, high-strength components such as turbocharger rotors and keeps machining to a minimum.

3.7   Slip casting

Slips are suspensions of one or more ceramic materials in a liquid, usually water, with a particle size around 1 μm and may be considered as colloidal systems (section 4.2). The slip casting technique involves pouring a slip into a porous mold often made from plaster of Paris (calcium sulphate hemihydrate) which absorbs liquid and deposits solid material at the mold walls. Excess slip is drained off after which the cast is removed and then fired. Slip casting is a very versatile technique and has been used for manufacture of tableware, sanitaryware, crucibles, tubes, thermocouple sheaths and gas turbine stators (Richerson, 1982). However, three factors affect the quality of a cast. First, particles must remain in suspension so that deposition occurs evenly on the mold walls. Secondly, high solid contents (*ca.* 70 weight %) improve the drainage rate and thirdly, low viscosity is required in order to prevent incorporation of air bubbles into the ceramic and to ease filling of the mold.

As for other colloidal dispersions the stability of slips is controlled by interaction forces between particles (section 4.3). Strong ceramics can be obtained in the absence of particle aggregates, which are avoided by using deflocculating agents in the slip. These materials, normally surface-active agents, probably act by adsorption at the solid–liquid interface, for both oxide and non-oxide ceramics, which modifies the interaction energy between particles (V_R in equation (4.9)). However, particle size, pH and ionic strength affect slip stability and rheological properties of these systems (Phelps & McLaren, 1978), important for controlling liquid drainage from the dispersions.

3.8   Injection molding

Injection molding is a plastic forming technique in which ceramic powder is added as a filler to an organic polymer, usually a thermoplastic, to form a plastically deformable mixture that is injected, by using a combination of heat and pressure, by a plunger into the mold. The viscosity before injection is an important parameter for controlling even filling of the mold and avoiding air bubbles in finished components. Large amounts of polymer are used, typically 30 volume %, although solid content in the mixture can be maximised by optimisation of the particle size, but binder removal is a difficult step as it can produce cracking and voids. Injection molding is used extensively in the plastics industry and has the potential for production of components with complex shapes such as turbine blades.

A second plastic-forming technique, extrusion, is related to injection molding although extrusion dies are open at one end in contrast with those used in the latter process. Extrusion is particularly useful when articles with long length and uniform cross-section are required. It has been applied to clay-based materials for manufacture of bricks, tiles and pipes although these systems exhibit the required plastic behaviour without polymer additives. However, other ceramic systems have been extruded, e.g. cordierite, $2MgO.2Al_2O_3.5SiO_2$ (Richerson, 1982), as monolithic honeycomb structures used for catalyst supports (Thomas, 1987) in the treatment of vehicle exhaust emissions.

Fabrication techniques described here do not constitute an exhaustive list and experimental conditions used for them are illustrated in the following chapters.

3.9   Strength and toughness in ceramic systems

Ceramic systems are strong under compression, a property used in steel reinforcement of concrete structures, but are brittle materials and weak in tension, which therefore limits their use. Engine components can undergo brittle fracture due to thermal shock during repeated cycling around their operating temperatures whereas ceramic substrates and dielectrics used in the electronics industry also experience similar failure because of thermal expansion mismatch between ceramic and metallic parts (Wiederhorn, 1984). Observed tensile strengths of ceramics are considerably lower than theoretical values. Thus, for normal glasses the latter are between 10 and 20 GPa, more than 100 times the experimental values (Holloway, 1986). The low strength of brittle solids was first explained by Griffith (1920) who considered growth of a crack with an elliptical cross-section through such a material. He showed that when the surface energy for formation of new fracture faces was balanced by release of elastic strain energy then the tensile fracture strength, S (also termed fracture strength) could be written as

$$S = (2E\gamma'/\pi C)^{1/2} \qquad (3.2)$$

where E is Young's modulus for the material, γ' the surface energy and $2C$ the crack length corresponding with the major elliptical axis.

29

The fracture toughness or stress intensity factor, K_{IC}, for a solid is defined by the expression

$$K_{IC} = (2E\gamma')^{1/2} \tag{3.3}$$

when the energy balance used for deriving equation (3.2) is applicable while values for S and K_{IC} are shown in table 1.3. Fracture occurs after crack initiation followed by growth until a critical size is reached, several tens of micrometres for ceramic materials. Applied stresses are concentrated up to 1000 times at the crack tip, sufficient for rupturing atomic bonds in the solid. A detailed analysis of fracture is outside the scope of this volume but accounts have been given in the book by Davidge (1979) and more recently by Wiederhorn (1984).

Fracture strength is a measure of the largest crack or flaw size in ceramic components. Various mechanisms introduce flaws (Lange, 1984); for example, diamond grinding can produce surface cracks. Secondly, ceramic powders are not usually handled at all stages under the clean-room conditions associated with the electronics industry so that voids, which also constitute flaws, arise when unintentional organic additives are removed on firing green bodies; inorganic impurities also act as sources of flaws. Fourthly, ceramic preparations such as powder mixing and coprecipitation which involve calcination of oxides tend to generate particle agglomerates. Different sintering rates between the latter and individual ceramic particles in a compact produce voids; the distinction between agglomerates and primary particles is explained in section 9.3. Dense, fine-grained ceramics are required for high strength, although flaws from different sources cause a distribution of strengths in supposedly identical components.

There has been increasing interest over the past decade in developing ceramic materials with not only enhanced strength and toughness but also with reproducible strength, an important requirement for exploitation of engineering ceramics. One approach involves synthesis of ideal powders for sintering. These powders are often considered to have a small particle size, a narrow range of sizes to avoid excessive grain growth, an absence of aggregates, sphericity for improved powder packing and a controlled chemical purity. Sub-micrometre mono-dispersed oxide spheres made by homogeneous nucleation of metal alkoxide solutions (section 5.7) fall into this category of powder. The spheres can pack to a structure containing pores of the same size so that a dense fine-grained ceramic forms at low temperature (table 5.4) when the green body is sintered. A different approach is physical treatment of a commercial powder such as Al_2O_3, obtained by precipitation in the Bayer process, in order to break down aggregates and reduce the flaw size. This pathway, sometimes termed colloidal processing is illustrated by results of Alford, Birchall & Kendall (1987). An aqueous colloidal dispersion (section 4.2) of alumina containing 40 weight % polyvinyl-alcohol-acetate was extruded through a narrow orifice under a high wall shear stress, 20 MPa. The fracture strength of sintered rods was 1.04 GPa, considerably smaller than 46 GPa, the theoretical tensile strength, but greater than the value (0.37 GPa) for powder not subjected to a shear stress. Another example of colloidal processing that involves fabrication of reaction-bonded silicon nitride by using Si powder made in a vapour-phase reaction is described in section 9.5.

Cracks do not always have a deleterious effect on the mechanical properties of ceramics. Zirconia exists in three crystallographic forms, and oxides such as MgO, Y_2O_3 and CaO are incorporated to stabilise the cubic phase. Pure zirconia remains in the tetragonal phase above 1473 K but undergoes a martensitic transformation on cooling to the monoclinic phase with an approximately 5% volume expansion. Two effects occur when tetragonal ZrO_2 is incorporated into a matrix. Firstly, the tetragonal$\rightarrow$monoclinic transformation produces radial microcracks around ZrO_2 particles on cooling when the monoclinic phase is added to a matrix such as cubic zirconia or alumina. An advancing crack is deflected and bifurcated by these microcracks, which increases the toughness of the material; this toughening mechanism is useful for improving thermal shock resistance of fabricated components. Secondly, particles in partially stabilised zirconia can be retained in the metastable tetragonal phase when the cubic matrix is cooled, if they have a small size, 1 μm or less. The high tensile stress at crack tips causes the tetragonal$\rightarrow$monoclinic transformation in the particles and this results in a compressive stress, which makes crack propagation difficult, thus increasing strength and toughness.

Toughening is also enhanced by using fibres to reinforce ceramic and metallic matrices. Composite materials are discussed further by Hull (1985) and the synthesis of fibres for application by matrix reinforcement is described in chapter 7. A recently developed technique for preparation of composites involves oxidation of molten metal (e.g. Al) containing fibrous filler (Newkirk *et al.*, 1987) whereby the ceramic matrix is

Ceramic fabrication

grown through the fibres. Unlike conventional processes, this Lanxide™ method avoids powder compaction so that grain boundary phases from additives are eliminated and the composite retains low-temperature strength at elevated temperature.

4 Sol–gel processing of colloids

4.1 Introduction

The phrase sol–gel describes several types of processes in different areas of chemistry and materials development, while the term 'gel' has been used (Flory, 1974) to embrace a wide range of substances in systems as diverse as lamellar mesophases, inorganic clays and oxides, phospholipids, disordered proteins and three-dimensional or network polymers. There are three types of sol–gel processes associated with corresponding transitions. Examples of the first are the reversible gelation of certain polysaccharide solutions, for example, agarose and the vulcanisation of rubber, but these transitions will not be considered in this book. The other two sol–gel processes have attracted intense interest since the mid-1970s because of their use in the synthesis of ceramic materials. Sol–gel processing of metal–organic compounds, namely, alkoxides, is described in chapter 5, while the early study of colloids, their stability and use in sol–gel transitions for both nuclear and industrial ceramics are the subject of this chapter.

4.2 The nature of colloids

The scientific study of colloids dates back to 1845 when Selmi prepared silver chloride dispersions (sols) followed by Prussian blue sols in 1847, which he referred to as demulsions and pseudosolutions, respectively. These systems were considered to be in the same category as starch, cholic acid and albumin, but distinct from true solutions, following experiments on the effect of salt on their sedimentation. Faraday (1857) observed light-scattering from ruby-coloured gold sols made by reduction of gold chloride with phosphorus together with the effect of salt on their stability and colour. However, the word colloid was first used (Graham, 1861) to describe 'glue-like' material prepared by dialysis of

Table 4.1. *Classification of colloidal systems*

System	Disperse phase	Dispersion medium
Dispersion (sol)	Solid	Liquid
Emulsion	Liquid	Liquid
Solid emulsion	Liquid	Solid
Foam	Gas	Liquid
Fog, mist or aerosol (of liquid particles)	Liquid	Gas
Smoke or aerosol (of solid particles)	Solid	Gas
Alloy, solid suspension	Solid	Solid

silicic acid made by acidifying silicate solutions and also organic species such as gums, caramel, tannin and albumin. Graham noticed that these colloids did not crystallise and had a lower diffusivity than molecular species implying a larger particle size.

Nowadays colloidal systems are defined as comprising a disperse phase with at least one dimension between 1 nm and 1 μm in a dispersion medium; examples of these systems are shown in table 4.1. For sols the colloidal dimension refers to particle diameter whereas for macroscopic colloidal systems such as foams it refers to film thickness.

The magnitude of this dimension distinguishes colloids from bulk systems in the following ways. First, they have very large interfacial surface areas, so that, for instance, dividing a 1 cm^3 cube of material into smaller cubes, each with volume 10^3 nm^3 increases the surface area from 6 cm^2 to 6×10^6 cm^2. Secondly, a significant percentage of molecules reside in the surface of colloidal systems such as sol particles. This is illustrated in figure 4.1, which shows the effect of length, l, of a cube on the surface to bulk distribution of molecules with volume 0.1 nm^3 and 0.02 nm^3 – values of l that lie within the colloidal dimension result in *ca.* 30% of molecules residing in the surface. These calculations also show that sol particles contain between 10^3 and 10^9 molecules.

4.3   The stability of colloids

Colloidal systems are usually described as lyophilic or lyophobic, and the former, which include macromolecular solutions such as polymers or proteins, together with micellar aggregates, are thermodynamically stable. However, lyophobic colloids (e.g. sols) are not thermo-dynamically stable as a consequence of the high interfacial area and their kinetic stability arises because of attractive forces and electrostatic repulsion acting between particles.

London showed with the advent of quantum mechanics that an attractive force could arise between two atoms or two molecules even when neither had a permanent dipole. The London interaction energy, V_L, between two identical atoms can be written as

$$V_L = -3h\,\nu_0\,\alpha^2/4r^6 \qquad (4.1)$$

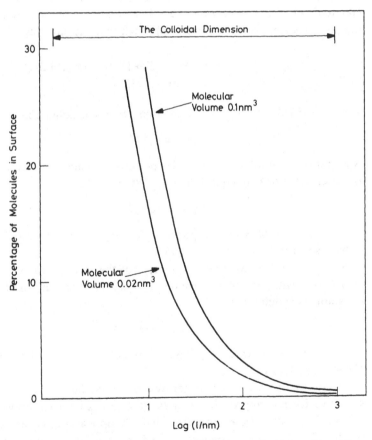

Figure 4.1. Percentage of molecules in the surface of a cube versus the logarithm of its length, *l*, for molecular volumes of 0.1 nm³ and 0.02 nm³.

where h is Planck's constant, α the static atomic polarizability, r the distance between atoms, and ν_0 the ground-state vibrational frequency for the electron. The interaction force $(\partial V_L/\partial r)$ was termed a dispersion force as ν_0 occurs in the theory of optical dispersion and $h\nu_0$ equated to the ionisation potential for the atom. The van der Waals energy is the sum of V_L and the potential due to dipole–dipole and dipole–induced dipole interactions. For example, the contributions of V_L to the total van der Waals energy for water, ethyl alcohol and carbon tetrachloride are 10.5%, 47.6% and 100% respectively (Hiemenz, 1986).

Based on the assumption that the van der Waals–London energy, V_A, between macroscopic bodies could be calculated if dispersion forces between individual atoms and molecules in the bodies were additive, Hamaker (1937) showed that the value of V_A for two equal spheres, radius a, separation H_0 at closest approach in a vacuum was

$$V_A = -A((x^2+2x)^{-1}+(x^2+2x+1)^{-1}+2 \ln (x^2+2x)/ \atop (x^2+2x+1))/12 \tag{4.2}$$

where $x = H_0/2a$ and A the Hamaker constant was defined as

$$A = \pi^2 q^2 \lambda^* \tag{4.3}$$

where q is the number of molecules per unit volume of material and λ^* the coefficient of r^6 in equation (4.1). When $x \ll 1$,

$$V_A = -Aa/12H_0 \tag{4.4}$$

The magnitude of V_A is responsible for aggregation of fine ceramic powders (1 μm or less in diameter) in air.

A composite Hamaker constant, A_{12}, for the interaction of two identical bodies with Hamaker constant A_1 in a liquid medium of Hamaker constant A_2 is defined as

$$A_{12} = (A_1^{1/2}-A_2^{1/2})^2 \tag{4.5}$$

and for oxides in water A_{12} has values between 0.5 and 5×10^{-20} J (Overbeek, 1982).

An alternative view of van der Waals forces, known as the macroscopic or Lifshitz approach, calculates the forces between bodies from knowledge of their bulk properties, namely optical properties over the whole electromagnetic spectrum, rather than considering one frequency, ν_0, as in the microscopic approach of Hamaker. It neither makes assumptions about the medium in which the interaction occurs nor whether dispersion forces between atoms are additive. This macroscopic approach has been described by Mahanty & Ninham (1976) and a review of attractive surface forces with reference to colloids is given by Tabor (1982).

Insoluble oxides in aqueous suspension develop surface electrical charges by surface hydroxylation followed by dissociation of surface hydroxyl groups as represented by the equations

$$M(OH)_{surface} + H_2O \rightleftharpoons MO^-_{surface} + H_3O^+ \qquad (4.6)$$

$$M(OH)_{surface} + H_2O \rightleftharpoons M(OH_2)^+_{surface} + OH^- \qquad (4.7)$$

The surface charge, ψ_0, a function of pH, results in compensating anions forming a diffuse charge layer around the surface. The region between the surface and undisturbed electrolyte in bulk solution over which ψ_0 decays to the bulk solution value is known as the electrical double layer, which has a reciprocal thickness κ given by the expression

$$\kappa = (2cNe^2z^2/\varepsilon_0\varepsilon kT)^{1/2} \qquad (4.8)$$

where c is the ionic strength of the solution, N the Avogadro number, e the electronic charge, z the valency of the ions, ε_0 the permittivity of the vacuum, ε the relative permittivity of the medium, k the Boltzmann constant and T the absolute temperature. In a 1–1 electrolyte such as potassium nitrate, κ^{-1} has a magnitude of *ca.* 100 nm when $c = 10^{-5}$ mol dm^{-3} and decreases to *ca.* 3 nm when $c = 10^{-2}$ mol dm^{-3}. A more detailed account of electrical double layers including effects of specific adsorption is given by Lyklema (1985).

Electrostatic repulsion occurs when electrical double layers of sol particles overlap. Double layers are at equilibrium if the particle approach is slow so that the energy of repulsion V_R can be equated, by using equilibrium thermodynamics, to the isothermal work for bringing particles from infinity to a separation of closest approach, H_0. For two flat plates in a z–z electrolyte

$$V_R = 64ckT(\tanh(ze\psi_0)/4kT)^2\exp(-\kappa H_0)/\kappa \qquad (4.9)$$

and a similar equation applies for spheres; specific adsorption can affect the choice of potential in equation (4.9).

An adsorbed layer of material, polymer or surface active agent, modifies the interaction of colloidal particles in several ways. When there is no overlap of layers V_A can be reduced (Vold, 1961), but an increase in free energy of the system, V_S, occurs on overlap. This has contributions from an entropy effect due to a limit on the conformation for layer material in the region of overlap, together with an osmotic pressure term for the same region. The stability brought about by adsorbed layers, referred to as steric stabilisation (Napper, 1982), is particularly important for non-aqueous systems when V_R can be negligible.

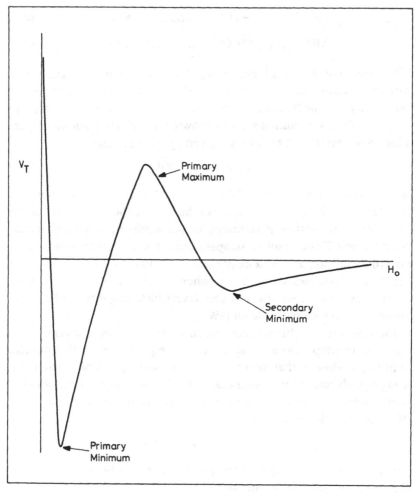

Figure 4.2. Schematic variation of total potential energy V_T versus separation at closest approach H_0 for two identical spheres.

Sol formation through cation hydrolysis

The stability of lyophobic colloids is nowadays considered in terms of the DLVO theory (Deryagin, Landau, Verwey & Overbeek) in which the total potential energy between two particles, V_T, is written as a sum of two components, V_R due to electrostatic double layer repulsion and the van der Waals energy, V_A (Verwey & Overbeek, 1948)

$$V_T = V_A + V_R \qquad (4.10)$$

In order to allow for other contributions such as V_S, V_T is written as

$$V_T = V_A + V_R + V_S \qquad (4.11)$$

and the variation of V_T with H_0 for two spheres (neglecting V_S) is shown in figure 4.2. Flocculation refers to reversible aggregation of sol particles in the secondary minimum at large H_0, whereas the primary maximum acts as a barrier to coagulation that is, irreversible aggregation in the primary minimum. Born repulsion occurs at very short separation.

4.4 Sol formation through cation hydrolysis

Many metal ions (M^{z+}) are subject to hydrolysis because of a high electronic charge (e.g. Zr^{4+}) or high charge density (e.g. Be^{2+}). Initial products of hydrolysis can condense and polymerise to form polyvalent metal or polynuclear ions, which are themselves colloidal (figure 4.3). For example, $(AlO_4Al_{12}(OH)_{24}(H_2O)_{12})^{7+}$ and $(AlO_4Al_{12}(OH)_{25}(H_2O)_{11})^{6+}$ are produced on increasing the pH (e.g. with sodium carbonate) of Al(III) salt solutions for mole ratios of $OH/Al \geq 2$. They have a structure consisting of twelve AlO_6 octahedra surrounding a single AlO_4 tetrahedron (Birchall 1983). Charge and pH determine whether H_2O, OH^- or O^{2-} act as ligands for the central cation as shown in figure

Example of a hydrolysis reaction

$M(H_2O)_b^{z+} \cdot \cdot [M(H_2O)_{b-1}OH]^{(z-1)\cdot} + H^\cdot$

Example of a condensation–polymerisation reaction

$2M(H_2O)_b^{z\cdot} \cdot \cdot [(H_2O)_{b-1}M(OH)_2M(H_2O)_{b-1}]^{(2z-2)\cdot} + 2H^\cdot$

Figure 4.3. Schematic representation of hydrolysis and condensation–polymerisation of polyvalent ions in aqueous solution.

4.4 (Kepert 1972) whereas the anion/cation mole ratio controls the degree of polymerisation and sol stability through its effect on V_R (equation (4.9)). Additional examples of polynuclear ions are shown in tables 4.2 and 4.3 (Baes & Mesmer, 1976), which illustrate the wide variety of cations that form hydrolysis products as well as the formation of different polynuclear species from a cation (table 4.3).

Hydrous metal oxides (also called hydroxides and hydrated metal oxides) are precipitated for values of $OH/M \geq z$ and can be converted to sols by peptisation. In this process adsorption of H^+ from dilute acid onto precipitate breaks up crystalline (or non-crystalline) aggregates into primary crystallites that disperse to a sol – the adsorption process probably modifies V_R in equation (4.9). When sols are made by peptisation the size of primary crystallites can be controlled by precipitation conditions such as temperature and reactant concentrations.

4.5   Outline of the sol–gel process for colloids

Sol–gel processing of colloids can be considered in the five stages shown in figure 4.5. The starting material, for example, reagent-grade metal salt, is converted in a chemical process to dispersible oxide, which forms

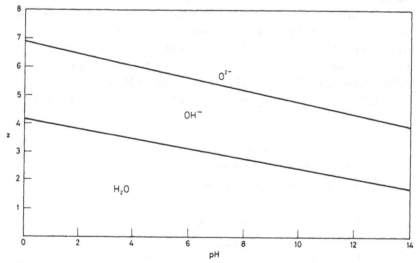

Figure 4.4. Effect of pH and valency, z, on formation of H_2O, OH^- and O^{2-} ligands (Kepert, 1972).

Table 4.2. *Occurrence of polynuclear cationic hydrolysis products (Baes & Mesmer, 1976)*

Species	Cation source
M_2OH^{3+}	Be^{2+}, Mn^{2+}, Co^{2+}, Ni^{2+}, Zn^{2+}, Cd^{2+}, Hg^{2+}, Pb^{2+}
$M_2(OH)_2^{(2z-2)+}$	Cu^{2+}, Sn^{2+}, UO_2^{2+}, NpO_2^{2+}, PuO_2^{2+}, VO^{2+}, Al^{3+}, Sc^{3+}, Ln^{3+}, Ti^{3+}, Cr^{3+}, Th^{4+}
$M_3(OH)_3^{3+}$	Be^{2+}, Hg^{2+}
$M_3(OH)_4^{(3z-4)+}$	Sn^{2+}, Pb^{2+}, Al^{3+}, Cr^{3+}, Fe^{3+}, In^{3+}
$M_3(OH)_5^{(3z-5)+}$	UO_2^{2+}, NpO_2^{2+}, PuO_2^{2+}, Sc^{3+}, Y^{3+}, Ln^{3+}
$M_4(OH)_4^{4+}$	Mg^{2+}, Co^{2+}, Ni^{2+}, Cd^{2+}, Pb^{2+}
$M_4(OH)_8^{8+}$	Zr^{4+}, Th^{4+}
$M_6(OH)_8^{4+}$	Be^{2+}, Pb^{2+}
$M_6(OH)_{12}^{6+}$	Bi^{3+}

Table 4.3. *Polynuclear anionic hydrolysis products of vanadium (V) at 298 K (Baes & Mesmer, 1976)*

Equilibrium

$$VO_2^+ + 2H_2O \rightleftarrows VO(OH)_3(aq) + H^+$$
$$10VO_2^+ + 8H_2O \rightleftarrows V_{10}O_{26}(OH)_2^{4-} + 14H^+$$
$$VO_2^+ + 2H_2O \rightleftarrows VO_2(OH)_2^- + 2H^+$$
$$2VO_2(OH)_2^- \rightleftarrows V_2O_6(OH)^{3-} + H^+ + H_2O$$
$$VO_3(OH)^{2-} \rightleftarrows VO_4^{3-} + H^+$$
$$2VO_3(OH)^{2-} \rightleftarrows V_2O_7^{4-} + H_2O$$
$$3VO_3(OH)^{2-} + 3H^+ \rightleftarrows V_3O_9^{3-} + 3H_2O$$

the sol on addition to dilute acid or H_2O. Removal of H_2O and/or anions from sol produces a stiff gel in the form of spheres, fibres, fragments or coatings and this transition is usually reversible. Calcination of gel in air yields oxide product after decomposition of salts whereas for non-oxide ceramics (carbides and nitrides) a carbon source is added at the sol stage and carbon-containing gel heated in a controlled gaseous atmosphere. For multicomponent oxides, sols are blended together before gelation and a component unavailable in sol form can be introduced as an electrolyte solution or oxide powder.

4.6 Sol–gel in the nuclear industry

During the late 1950s the nuclear industry was particularly interested in mixed $(Th,U)O_2$ fuels for thermal reactors because of the larger abundance of thorium in the Earth's crust than uranium. In this fuel cycle non-fissile ^{232}Th was converted to fissile ^{233}U through the reaction,

$$^{232}Th \xrightarrow{n\gamma} {^{233}Th} \xrightarrow[t_{1/2} = 23.5 \text{ min}]{\beta} {^{233}Pa} \xrightarrow[t_{1/2} = 27.4 \text{ days}]{\beta} {^{233}U} \qquad (4.12)$$

where $t_{1/2}$ refers to half-life. The decay of ^{232}U associated with ^{233}U gave rise to daughter products, two of which, ^{208}Tl and ^{212}Bi, were emitters of high-energy γ-rays, 2.6 MeV for the former and 0.8–2.2 MeV for ^{212}Bi (Arnold, 1962). This necessitated fuel fabrication and reprocessing to be done in a remote handling facility in order to minimise radiation doses to operators. It was very important to avoid coatings of radioactive particles on equipment and inside glove boxes so that the existing fabrication route to pelleted fuel was eliminated because of its ability to generate dust.

Initial work at Oak Ridge National Laboratory, ORNL, in the USA showed that HNO_3 could be removed from hydrated thorium nitrate (i.e. the starting material in figure 4.5) by using superheated steam at temperatures up to 748 K (Dean *et al.*, 1962; Ferguson, Dean & Douglas, 1964). The residue (NO_3/Th mole ratio 0.03) was the dispersible oxide and formed a ThO_2 sol in dilute HNO_3 (NO_3/Th = 0.1) or uranyl nitrate solution with a typical oxide composition $(U_{0.03}Th_{0.97})O_2$. Sol could be reversibly tray-dried at 363 K to gel fragments, density 5000 kg m^{-3} (55% theoretical density, TD), which contained 5 weight percent H_2O. A homogeneous solid solution with 98% TD resulted when gel was calcined at 1423 K in air and then in 4% H_2/Ar (to reduce UO_3 to UO_2) a significantly lower temperature, compared with 2000 K required in conventional fabrication techniques, to achieve densification.

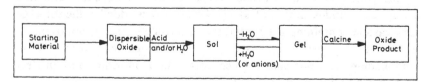

Figure 4.5. Outline of the sol–gel process for colloids (Woodhead & Segal, 1984).

One application of dense ThO_2–UO_2 ceramics was for high-temperature reactors (HTR) where spherical fuel powders were embedded in a graphite matrix; pyrolytic carbon coatings on the spheres acted as containment for both gaseous and solid fission products. In addition, pressurised water reactor (PWR) fuel pins could be filled easily with spheres by vibrocompaction. Spherical powders were made by dispersing concentrated thoria sol (5 M in Th) to an emulsion in an immiscible organic solvent capable of extracting H_2O from sol, for example, 2-ethyl hexanol (Ferguson *et al.*, 1964) and gelation occurred during this dehydration process. Haas & Clinton (1966) showed that small quantities (*ca.* 0.5 volume %) of the surfactant sorbitan mono-oleate (Span 80) in the solvent prevented coalescence of liquid drops and improved gel sphericity by modifying surface tension at the liquid–liquid interface. Sphere size was controlled by sol and surfactant concentrations together with stirring rate, which enabled mixed oxides to be prepared on the 1000 kg scale with diameters between 50 and 2000 μm.

Alternative methods for sol preparation and gelation were developed at other centres with nuclear expertise. At the Harwell Laboratory in the United Kingdom, Hardy & co-workers (1968) made thoria sols (5 M in Th, NO_3/Th = 0.08) from dispersible oxide obtained by thermal denitration of hydrated thorium nitrate at 763 K in air; oxide porosity was controlled by adding carbon to sol and then burning it out in the calcination step (figure 4.5). Carbon was also used for preparing mixed Th/U dicarbides (Hardy, 1968; Kelly *et al.*, 1965) by carbothermic reduction of carbon-containing gel at 1873 K in vacuum, compared with 2373 K required for conventional methods; this illustrates, as for oxides, the intimate mixing achieved by using colloidal intermediates. This lower temperature allowed the use of standard furnaces with high reliability in remote operations and avoided grinding hard actinide dicarbides.

Internal gelation was pioneered in the Netherlands (Hermans & Slooten, 1964; Hermans, 1968) at KEMA (Noalaze Vennaotschap tot Keuring van Electrotenische, Materialen, Arnhem). An emulsion of sol droplets in trichloroethylene could be gelled by diffusion of $NH_3(g)$ through the solvent. However, a skin was formed on liquid drops during this external gelation process that produced cracks in spheres because of osmotic pressure effects: the maximum attainable sphere diameter was 25 μm. Preparation of crack-free spheres 100–700 μm diameter was achieved by using internal gelation that involved addition of an ammonia donor such as urea, $CO(NH_2)_2$, or hexamethylenetetramine, $(CH_2)_6N_4$

to the sol before emulsification; $NH_3(g)$ released on warming solvent to 363 K caused gelation.

The use of long-chain amines to remove hydrolytic nitric acid from uranyl nitrate solutions was carried out in Italy at CNEN (Comitatio Nazionale perl'Energia Nucleare). A solution of the primary amine, Primene JMT, $R_3–C–NH_2$ ($C_{18} < R < C_{22}$) was mixed with 2 M uranyl nitrate solution (Cogliatti *et al.*, 1964), resulting in a denitrated solution after separating the liquid phases, with NO_3/U between 1 and 1.7, that was concentrated to 3.5 M in U(VI) and then catalytically reduced to U(IV). The method could be applied to both sol and sphere preparation, the latter by external gelation in which sol droplets were gelled on removal of anions when Primene JMT was added to the solvent. Recent developments in the application of sol–gel processing to ceramic nuclear fuels have been described by Turner (1986) and Ganguly *et al.* (1986).

4.7   Gel precipitation in the nuclear industry

Gel precipitation is a process that combines precipitation and gelation in one step, does not require sol as starting material and together with sol–gel is sometimes referred to as a gel-processing technique. Contributions to its development have been made by the United Kingdom Atomic Energy Authority and AGIP Nucleare S.p.A, Milan, where it was referred to as the SNAM process (Brambilla, Gerontopoulos & Neri, 1970); an outline of the British route is shown in figure 4.6. A feed solution of plutonium and uranyl nitrates containing an organic gelling aid and modifying agent (for stabilisation) was dispersed through a vibrating orifice to droplets, which fell into aqueous ammonia where they formed mixed hydroxide gel spheres. After drying, spheres were debonded in CO_2 for removal of organics and sintered to dense oxides at 1723 K in 5% H_2/Ar. The method was suitable for loading fast-reactor fuel pins by vibrocompaction using two sphere diameters, 80 μm and 800 μm, with the absence of radioactive dusts. The quality of spheres that can be made by using this technique is shown in figure 4.7.

4.8   Industrial applications of sol–gel processing

The development of sol–gel ceramic nuclear fuels showed that stable concentrated dispersions could be made in multikilogram quantities by dust-free routes and converted to spherical oxide powders with both homogeneous composition and controlled particle size. These process properties together with achievement of theoretical density at relatively low temperatures with an associated energy economy are attractive features for industrial ceramics particularly if no cost penalties are incurred during preparation.

At Oak Ridge a systematic study was made by Hardy & co-workers (Hardy, Buxton & Lloyd, 1967a; Hardy, Buxton & Willmarth, 1967b) on lanthanide hydroxide sols made by peptising freshly prepared and aged hydroxide precipitates with HNO_3; a representative dispersion was

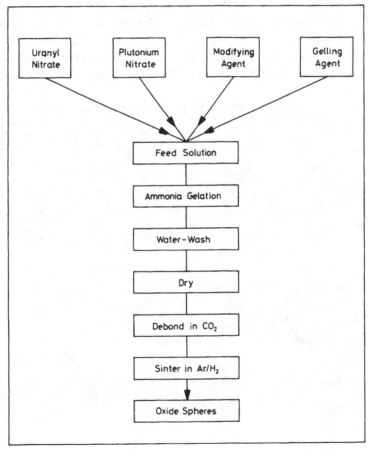

Figure 4.6. Outline of the gel precipitation route to mixed UO_2–PuO_2 spheres (Marples *et al.*, 1981).

2.3 M in Sm, had pH 7.0 and a NO_3/Sm mole ratio of 0.19. Transmission electron micrographs showed that initial precipitates (e.g. $Pr(OH)_3$) consisted of amorphous spheres 3–6 nm diameter which aged to rod-like crystals 30–60 μm long that aggregated to bundles. Microspheres, 20–200 μm diameter, made by dehydration contained randomly orientated crystal bundles and sintered to dense powders, e.g. 97.5% TD for Nd_2O_3 at 1373 K.

Processes were developed at Harwell for sols of zirconia (Woodhead, 1970), ceria (Woodhead, 1974), indium oxide (Bones & Woodhead, 1974) and titania (Woodhead, 1975). The peptisation process is illustrated for ceria sol preparation. Transmission electron micrographs of unconditioned Ce(IV) hydrate precipitate made by adding a NH_4OH/H_2O_2 mixture to Ce(III) nitrate solution indicated aggregates of 'beads' *ca.* 0.1 μm diameter (figure 4.8) although X-ray diffraction showed the primary particle size was *ca.* 8 nm (Woodhead & Segal, 1984). After conditioning with HNO_3, the precipitate dispersed to sol

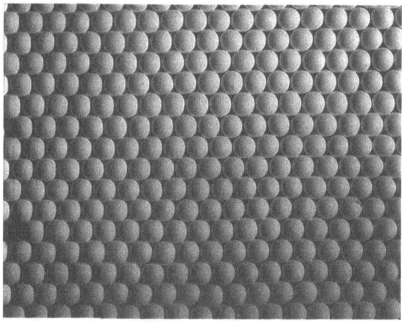

2000 μm

Figure 4.7. Photomicrograph of gel precipitated alumina spheres (Courtesy of the UKAEA.)

Table 4.4. *Effect of processing conditions on the density of sol–gel ceria (Nelson et al., 1981)*

HNO_3/CeO_2 mole ratio	Gel density[a] $(kg\ m^{-3})$	Oxide density[b] $(kg\ m^{-3})$
0.00	2000	2300
0.32	2400	3000
0.53	3600	5400
1.00	4000	6200

[a] After drying at 378 K.
[b] After calcination at 1273 K; theoretical density of ceria = 7100 kg m^{-3}.

with particle size 8 nm (figure 4.8*b*, *c*), which was tray-dried to gel fragments (figure 4.9) with the same X-ray line broadening as unconditioned precipitate. The NO_3/CeO_2 ratio affects the degree of deaggregation and this is reflected in gel and oxide density (table 4.4) so that deaggregated hydrate yields dense CeO_2 at 1273 K. As for ThO_2 sols, high density obtained with this non-aggregated densifiable sol arises from close packing of primary crystallites. Other dispersions can be prepared that give porous materials on conversion to gels and oxides. These non-densifiable or aggregated sols (Nelson *et al.*, 1981) consist of colloidal units that contain aggregates of primary particles. Densifiable and non-densifiable TiO_2 sols are shown in figure 4.10 and can be distinguished by

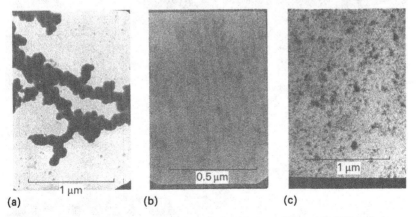

(a) (b) (c)

Figure 4.8. Transmission electron micrographs for (a) unconditioned Ce(IV) hydrate precipitate, (b) and (c) conditioned precipitate (Courtesy of the UKAEA.)

47

Table 4.5. *Densification of titania prepared from non-aggregated and aggregated sols and conventional precipitates (Nelson et al., 1981)*

Temperature of calcination/K	Density/(kg m^{-3})^a		
	Aggregated	Non-aggregated	Precipitate
Dried only	1490	2590	—
673	1500	3250	2340
873	1590	3450	—
1073	2640	4070	2910
1273	3730	4110	—

^{a} Theoretical density of anatase 3850 kg m^{-3}; theoretical density of rutile 4260 kg m^{-3}.

their different light-scattering properties, while the effect of colloid structure on oxide density is shown in table 4.5. Recent studies on the characterisation of oxide sols by neutron scattering techniques have been described by Ramsay (1986).

Figure 4.9. Ceria sol and tray-dried gel (Courtesy of the UKAEA.)

Examples of porous oxides made from aqueous sols are spherical powders for chromatographic adsorbents (Ramsay, 1978) and catalyst supports (Cairns, Segal & Woodhead, 1984). Other free-flowing spherical powders include electrically conducting ceramics, 3 % SnO_2–In_2O_3 and cubic ferrites, $Ni_{0.3}Zn_{0.7}Fe_2O_4$ (Woodhead & Segal, 1985), Y_2O_3-stabilised zirconia for engineering ceramic applications (Woodhead, 1984) and crystalline ceramics (Synroc) for solidification of high-level radioactive liquid waste. Synroc, an abbreviation for synthetic rock, is based on naturally occurring minerals, hollandite ($BaAl_2Ti_6O_{16}$), perovskite ($CaTiO_3$) and zirconolite ($CaZrTi_2O_7$) and has been proposed (Reeve & Ringwood, 1983) as an alternative to glass for waste storage because of possible improved leach resistance. Although it can be prepared by hot-pressing appropriate oxides, sol–gel Synroc powders (5–25 μm diameters) flow easily and have been isostatically hot-pressed under 413 MPa at 1598 K in air to high-density discs (4300 kg m^{-3}, Segal & Woodhead, 1986). Spray-drying, an alternative technique for converting sol to spherical gel powders, is more economical for multi-kilogram oxide preparations than chemical gelation using solvents. An

Figure 4.10. Aggregated and non-aggregated titania sols. (Courtesy of the UKAEA.)

49

example of spray-dried powder is shown in figure 4.11. Aqueous sols are not, in the author's experience, suitable materials for conversion to sub-micrometre spherical powders by the gelation techniques described in this chapter.

A sol–gel process that has been taken to the manufacturing stage concerns alumina-based abrasive grain used in grinding wheels and coated papers. The conventional route (figure 4.12) involves fusion of Al_2O_3 with dopant, usually ZrO_2 or MgO, at temperatures over 2073 K and allowing melt to solidify. However, variations in cooling rate result in a variable product with a wide range of crystallite sizes of one phase in a matrix of the second phase. The solid abrasive is comminuted by extensive grinding and grain that is unsuitable for use because of its size is recycled by fusion. In the sol–gel route (figure 4.12), an aqueous pseudoboehmite (α-Al_2O_3.H_2O) sol is doped with a second component introduced as oxide powder or salt solution that causes gelation (Leitheiser & Sowman, 1982). The brittle gel, which can be crushed and sieved, easily transforms to a continuous phase of randomly orientated α-Al_2O_3 crystallites containing a secondary phase, for example $MgAl_2O_4$ after sintering at 1623 K. The crystal size of *ca.* 300 nm is considerably

20 μm

Figure 4.11. Spray-dried alumina. (Courtesy of the UKAEA.)

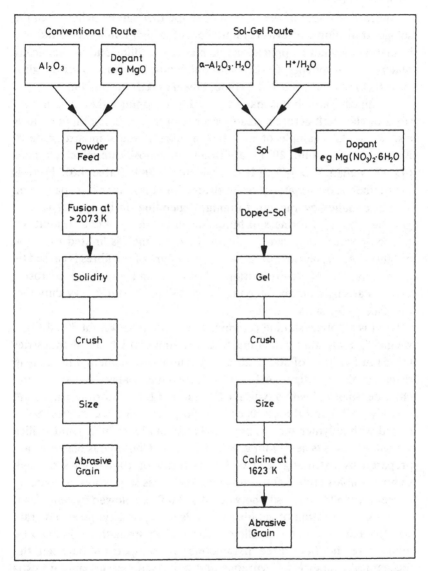

Figure 4.12. Comparison of conventional with sol–gel route to alumina-based abrasive grain.

smaller than in fused grain and imparts improved abrasive properties to the sol–gel material.

Another example of the control of shape that can be achieved in the sol–gel transition concerns ceramic fibres for high-temperature thermal insulation in general applications such as cavity filling and in chemically reactive environments, for example wall linings in furnaces for the glass and steel industries (Griffiths, 1986; Sowman & Johnson, 1985).

Aluminosilicate fibre is manufactured by dropping either an alumina–silica or clay melt across a steam or air jet although spinning of melts is also used. An example of this 'blown' fibre is the standard grade of Kaowool (Morganite, 1981), a vitreous aluminosilicate which contains (43–47) weight % Al_2O_3; (57–53) weight % SiO_2 (table 4.6). Fibre is used either as bulk material or fabricated into mats, blankets, paper and complex shapes by vacuum forming, depending on its application. In general, the refractoriness of fibres increases with alumina content but the composition that can be obtained from a melt is limited to *ca.* 50 weight % Al_2O_3 because of its viscosity. Failure of the fibres is caused by a combination of devitrification, shrinkage and attack by corrosive vapours arising from metallic impurities (V and Na) in heavy-duty fuel oils used in furnaces.

Saffil is a polycrystalline ceramic, composition 95 weight % Al_2O_3; 5 weight % SiO_2, that is manufactured by Imperial Chemical Industries (1975) in the form of staple fibre and continuous filament with a mean diameter of 3 μm (figure 4.13). The starting material is basic aluminium chloride 'solution' with a typical Cl/Al ratio of 1.1:1 (Morton, Birchall & Cassidy, 1974) and these sols contain polymeric cationic species. Sol is doped with polymer such as polyvinyl alcohol (2 weight %) and a silica source, which acts as a grain growth inhibitor. Continuous filaments are prepared by extruding feed with a viscosity of 10–100 Pa s through spinneret holes (100–200 μm diameter) whereas staple fibre is made by concentration to a viscosity between 0.5–2.0 Pa s followed by centrifugal spinning. Calcination of gel fibre controls total porosity, pore and grain size together with the crystalline phases. High strength is obtained by minimising the first three parameters. A comparison between the mechanical properties of vitreous and polycrystalline ceramic fibres is shown in table 4.6.

Sol–gel ceramics described in the preceding paragraphs have been derived from trivalent and tetravalent cationic sources. Examples of sol–gel products obtained from a divalent cationic source are varistors, based on zinc oxide, which are electrical resistors that do not obey Ohm's law and have applications as surge arrestors and voltage limiters in electrical devices. The conventional synthesis involves sintering powder mixtures of ZnO with additives (e.g. Bi_2O_3) at 1473–1673 K, which results in large grain sizes, 7–13 μm. Lauf & Bond (1984) prepared sols of zinc hydroxide and dopants before combining them and drying the mixed dispersion at 373 K to glassy gel fragments with a typical oxide composition 0.5 mole % Bi_2O_3; 0.5 mole % CoO; 0.5 mole % Cr_2O_3; 0.5 mole % MnO; 1.0 mole % Sb_2O_3; 97.0 mole % ZnO. Hot-pressing in vacuum at 963–1073 K followed by heating in air at 973–1273 K produced a material with non-ohmic electrical properties and smaller grain size (3–4 μm) compared with conventional varistor material, which can lead to more compact devices.

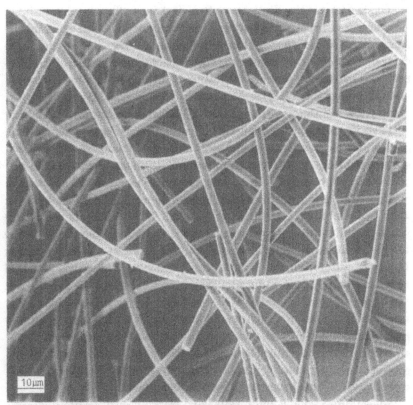

Figure 4.13. Scanning electron micrograph for Saffil alumina fibre. (Courtesy of Imperial Chemical Industries.)

Table 4.6. *Mechanical properties of vitreous and polycrystalline ceramic fibres*

Property	Saffil alumina	Kaowool
Crystallinity	Polycrystalline	Vitreous
Composition	95 weight % Al_2O_3	(43–47) weight % Al_2O_3
	5 weight % SiO_2	(57–53) weight % SiO_2
Density/(kg m^{-3})	2500	2560
Melting point/K	2253	2033
Maximum usable temperature/K	1673	1533
Tensile strength/GPa	1.0	1.4
Young's modulus/GPa	100	120

Oxide coatings also illustrate control of shape in the sol–gel transition. Advantages of applying a coating from the liquid phase are its simplicity and cost, but cracking and loss of adhesion can occur on drying. Thin craze-free layers (*ca.* 1 μm) for improved oxidation resistance of 18%Cr–8%Ni–Ti-stabilised stainless steel were deposited by dipping wire mesh into non-aggregated CeO_2 and SiO_2 sols (Nelson *et al.*, 1981), drying the liquid film and sintering the gel at 1073 K, whereas porous coatings for catalytic applications were obtained by using an aggregated alumina sol. Silicon wafers have been coated with SiO_2 by spinning or dipping substrate into silica sol (derived from a fumed or flame-hydrolysed oxide) combined with acrylic latex, a copolymer formulation of methylmethacrylate and 2-ethylhexylacrylate containing 2 mole % methacrylic acid (Yarborough, Gurujava & Cross, 1987). Adherent crack-free coatings with thicknesses between 5 and 50 μm were obtained after firing at 1073 K although polyethylene glycol was added to sol for the thicker layers. Coatings ranging from tens of micrometres to several hundred micrometres in thickness for applications in wear resistance can be built up by plasma-spraying (Scott & Cross, 1986). In this process the coating material is melted in a high-temperature (*ca.* 5000 K) plasma. Molten particles are then propelled at high velocity onto the surface of a substrate where they are quenched and adhere to form the coating. Sol–gel powders are particularly useful in plasma-spraying because of their sphericity and flow properties (Scott & Woodhead, 1982).

Sol–gel processing is often associated with non-melt routes to glasses that involve hydrolysis of silicon alkoxides, mixed alkoxide–colloid systems and gelation of silica sols. Some colloidal silicas are available commercially under the trade names of Ludox (Du Pont) and Syton (Monsanto). These alkaline dispersions can be made by an ion-exchange process in sodium silicate solutions, a technique that has been used (Livage & Lemerle, 1982) for sols and gels of V_2O_5 where decavanadic acid, obtained by treatment of sodium metavanadate with resin poly-merised over a period of hours to dark red gels.

Early studies (Roy, 1969) showed that colloidal SiO_2 containing solu-tions of carbonates, hydroxides or nitrates of alkali metals, alkaline-earths, lanthanides, Ga, Fe, Ti, Zr, Th and Pb set to rigid gels over a period of hours. Gels were dehydrated, heated at 873–1273 K to de-compose salts and pressed to pellets that were transformed to solid glass fragments of homogeneous composition by sintering at 1473–1673 K. Rabinovich & co-workers (1982, 1984) developed a 'double' processing method for making large crack-free tubes (25 cm long, 2.3 cm external diameter, 1.7 cm internal diameter) that exhibited small shrinkages in the gel state. Fumed silica sol was dried, calcined at 1173 K and then re-dispersed for SiO_2/H_2O weight ratios between 0.55 and 0.62, a process considered to result in aggregates larger than in the original sol. Linear shrinkage (2–4%) was minimised because of limited contacts between aggregates and a modified pore-size distribution in the gel encouraged rapid diffusion of water from bulk material with a reduction of non-uniform stresses responsible for cracking. Non-transparent, translucent and optically transparent silica glasses were obtained by sintering gels at 1523–1573 K, at 1723 K in He or air and at 1723–1743 K in He res-pectively. For boron-containing glass, up to 4 weight % B_2O_3 was intro-duced by doping sol or impregnating gel with boric acid solution. Densification was enhanced by the low viscosity of B_2O_3, high reactivity towards fine SiO_2 particles and reduction in crystallisation rate of SiO_2 because of the presence of B_2O_3.

An alternative approach to gelation was taken by Scherer & Luong (1984) in which a fumed silica powder with particle size 45 nm was dis-persed in chloroform and sterically stabilised by an adsorbed layer of *n*-decanol. Electrostatic repulsion was absent in this dispersion as addition of up to 10^{-2} M (n–C_4H_9)$_4$NBF$_4$ did not affect sol stability. Sols were gelled by deprotonation of silanol groups brought about on addition of $NH_3(g)$ or amines at low concentration (10^{-6}–10^{-3} M). Smaller capillary pressures (equation (5.12)) are produced on evapora-tion of organic solvent compared with water because of lower surface

Table 4.7. *Critical parameters for water and non-aqueous solvents*
(Bonner et al., 1985)

Solvent	Boiling point K	Critical pressure MPa	Critical temperature K
Acetone	329	4.7	508
Methanol	336	8.0	513
Ethanol	351	6.4	513
1-Propanol	370	5.2	538
1-Butanol	390	4.4	563
Ethylene glycol	470	7.7	643
Water	373	21.9	648

tensions and gels were sintered to glass rods (7 cm long, 1.3 cm diameter) at 1723 K in a He/Cl$_2$ atmosphere. Non-aqueous systems allow control of the water content for gelation together with removal of solvent water under supercritical conditions (table 4.7; Bonner, Kordas & Kinser, 1985), and high-density glass has been obtained from both types of sol (Chandrashekhar & Shafer, 1986). Gels derived from aqueous and ethanolic fumed silica dispersions sintered at 1593 K and 1573 K, respectively, to glasses with densities of 2250 kg m^{-3}, the same as for fused silica glass.

Sol–gel processing of colloids has also been applied to non-oxide ceramics of industrial interest. Sols derived from flame-hydrolysed SiO$_2$ were doped with C powder and either spray-dried to spherical powders (diameter *ca.* 20 μm) or tray-dried to fragments (Szweda, Hendry & Jack, 1981). For spray-dried gel, 100% sub-micrometre α-Si$_3$N$_4$ was obtained at 1673–1773 K in a N$_2$/H$_2$ atmosphere when the C/SiO$_2$ mole ratio was more than 2:1, whereas β-Si$_3$N$_4$ formed at 1837 K. Oven-dried gel fragments gave α-Si$_3$N$_4$ between 1673 and 1773 K for mole ratios of 2:1 because their water content, lower than for spray-dried gel, did not cause oxidation of carbon.

4.9   Summary

Aqueous and non-aqueous oxide sols have been widely investigated for the synthesis of glasses, crystalline oxides and non-oxide materials. Sol–gel processing allows fabrication of ceramics as spherical powders,

Industrial applications of sol–gel processing

fibres, coatings or monoliths by direct sintering of gels and in some cases manufacturing routes have been developed. The technology is versatile because it uses mainly aqueous dispersions, is energy efficient because of reactivity at lower temperatures compared with syntheses involving fusion or sintering powders and is suited for compositions that are difficult to make by conventional methods. Colloidal processing can result in homogeneous chemical compositions and allows control of oxide porosity.

5 Sol–gel processing of metal–organic compounds

5.1 Introduction

Sol–gel processing of colloids described in chapter 4 allows the preparation of homogeneous compositions and crystalline phases at temperatures lower than required for conventional powder mixing. However, sol particles contain between 10^3 and 10^9 molecules and improved mixing of components can occur by interaction at the molecular rather than colloidal level. This is achieved during the hydrolysis of metal–organic compounds, namely alkoxides. Sol–gel processing of alkoxides has attracted intense interest in the past 15 years because it offers non-melt routes to high-purity glasses and crystalline ceramics. The synthesis and hydrolytic reactions of alkoxides are described in this chapter, together with their use in the preparation of a wide range of ceramic materials.

5.2 The synthesis of metal alkoxides

Metal alkoxides have the general formula $M(OR)_z$ and can be considered to be derivatives of either an alcohol, ROH where R is an alkyl chain, in which the hydroxyl proton is replaced by a metal M, or of a metal hydroxide, $M(OH)_z$.

Ebelman (1846) made the first synthesis of an alkoxide, silicon tetra-isoamyloxide, by reacting silicon tetrachloride with isoamyl alcohol. This was followed (Ebelman & Bouquet, 1846) by the preparation of boron methoxide, ethoxide and amyloxide using boron trichloride and the corresponding alcohol. Electronegativity of the main element is an important factor in the choice of synthetic route so that strongly electropositve metals with valencies up to three react directly with alcohols to give alkoxides together with the liberation of hydrogen. For example, alkali metals (Ethyl Corporation, 1955) and alkaline earths (Ca, Sr and Ba) give alkoxides directly (Smith II, Dolloff & Mazdiyasni, 1970), but

The synthesis of metal alkoxides

aluminium (Teichner, 1953) requires a catalyst, I_2 or Hg. This direct route is also used for lanthanides and yttrium (Mazdiyasni, Lynch & Smith II, 1966) e.g.

$$2Y + 6C_3H_7{}^iOH \xrightarrow{HgCl_2} 2Y(O^iC_3H_7)_3 + 3H_2 \qquad (5.1)$$

where the $HgCl_2/Y$ mole ratio is 10^{-3}–10^{-4}. (In the description of alkoxides, the superscripts n, t, s, and i refer to normal, tertiary, and secondary (or iso) alkyl chains, respectively.)

The reaction of alcohols with metal halides to form alkoxides is particularly useful for electronegative elements, for example, silicon (Bradley, Mehrotra & Wardlaw, 1952)

$$SiCl_4 + 4C_3H_7{}^iOH \rightleftarrows Si(O^iC_3H_7)_4 + 4HCl \qquad (5.2)$$

The reactivity for a chloride within a particular group of the Periodic Table decreases with increasing electropositive character of the metal. Hence the reaction (Bradley, Saad & Wardlaw, 1954) of C_2H_5OH with chlorides of Ti, Zr and Th, can be considered to give $TiCl_2(OC_2H_5)_2.$ C_2H_5OH, $\{ZrCl_2(OC_2H_5)_2.C_2H_5OH$, and $ZrCl_3(OC_2H_5).C_2H_5OH\}$ and $ThCl_4.4C_2H_5OH$ as the products. However, for most metals the reaction can be forced to completion by using bases. Ammonia was used in the case of tantalum (Bradley, Wardlaw & Whitley, 1956), whereas the pyridine route has been used for plutonium alkoxides (Bradley, Harder & Hudswell, 1957).

Non-metallic oxides and hydroxides react with alcohols to form esters (i.e. alkoxides of non-metallic elements) and water

$$M(OH)_z + zROH \rightleftarrows M(OR)_z + zH_2O \qquad (5.3)$$

$$MO_z + 2zROH \rightleftarrows M(OR)_{2z} + zH_2O \qquad (5.4)$$

These reactions are reversible so that water has to be removed continually in order to prevent ester hydrolysis; this can be achieved by using solvents such as xylene in which azeotropes formed with water are fractionated out. Boron alkoxides, $B(OR)_3$ have been synthesised by dehydrating boric acid or oxide with alcohol, and vanadyl alkoxides have been prepared (Mehrotra & Mittal, 1964) by the reaction of oxide with alcohols in benzene.

Alcohol interchange or alcoholysis reactions can be represented as

$$M(OR)_z + bR'OH \rightleftarrows M(OR)_{z-b}(OR')_b + bROH \qquad (5.5)$$

and involve replacement of one alkoxy group, OR with another, OR'. These reactions have been applied widely to various elements for example, antimony (Mehrotra & Bhatnagar, 1965), and aluminium (Mehrotra, 1953)

$$Al(O^iC_3H_7)_3 + 3C_4H_9{}^nOH \rightarrow Al(O^nC_4H_9)_3 + 3C_3H_7{}^iOH \quad (5.6)$$

where isopropanol, $C_3H_7{}^iOH$, can be fractionated out and niobium (Bradley *et al.*, 1958)

$$Nb(OC_2H_5)_5 + 4C_3H_7{}^iOH \rightarrow Nb(OC_2H_5)(O^iC_3H_7)_4 + \\ + 4C_2H_5OH \quad (5.7)$$

In transesterification reactions an equilibrium is established between a metal alkoxide and ester CH_3COOR'

$$M(OR)_z + bCH_3COOR' \rightleftarrows M(OR)_{z-b}(OR')_b + bCH_3COOR \quad (5.8)$$

Distillation of the more volatile ester in equation (5.8) generates new alkoxide so that for zirconium *tert*-butoxide (Mehrotra, 1954) isopropyl acetate is fractionated out

$$Zr(O^iC_3H_7)_4 + 4CH_3COO^iC_4H_9 \rightarrow Zr(O^iC_4H_9)_4 + \\ + 4CH_3COO^iC_3H_7 \quad (5.9)$$

The final method that will be mentioned for alkoxide synthesis concerns oxidation of metal alkyls

$$MR_z + z/2\ O_2 \rightarrow M(OR)_z \quad (5.10)$$

This reaction is used to manufacture aluminium alkoxides containing even-numbered linear alkyl chains from ethylene, hydrogen and aluminium feedstocks (Condea Chemie, GmBH, 1986). An excellent account of synthetic pathways to alkoxides is given in the book by Bradley, Mehrotra & Gaur (1978), and review articles on preparative routes have been made by Mehrotra (1983) and later by Okamura & Bowen (1986).

5.3   *The physical properties of alkoxides*

The electronegativity of elements affects both the synthetic route to alkoxides and their physical properties. Hence sodium alkoxides are ionic solids, Ge, Al, Si, Ti and Zr alkoxides are often covalent liquids whereas lanthanide and yttrium alkoxides with intermediate electronegativities are mainly solids (Mazdiyasni, 1982); the covalent nature of

The physical properties of alkoxides

Table 5.1. *The volatility of representative alkoxides (Bradley et al., 1978)*

Alkoxide	Boiling point or sublimation temperature at reduced pressure/K
$Si(OC_2H_5)_4$	441 at atmospheric pressure
$B(O^nC_4H_9)_3$	401 at atmospheric pressure
$Al(O^sC_4H_9)_3$	445 at 67 Pa
$Fe(OC_2H_5)_3$	428 at 13 Pa
$Ti(O^iC_3H_7)_4$	322 at 13 Pa

Table 5.2. *Effect of distillation on the purity of tetraethoxysilane, TEOS (Gossink et al., 1975)*

TEOS	Impurity (p.p.b.)					
	Mn	Cr	Fe	Co	Ni	Cu
As supplied	10	15	86	0.7	<200	<200
Once distilled	0.8	2	31	0.3	11	<20

Table 5.3. *Chemical purity of a flame-hydrolysed silica powder, Aerosil OX50, (Degussa Company, 1981)*

Impurity	Concentration (p.p.m.)
Al_2O_3	<800
Fe_2O_3	<100
TiO_2	<300
HCl	<100

alkoxides increases with the degree of alkyl-chain branching. The fact that many alkoxides are liquids or volatile solids (table 5.1) means that they can be purified by distillation (Gossink *et al.*, 1975) to form exceptionally pure oxide sources (table 5.2). This level of purity may be compared with impurity levels in flame-hydrolysed SiO_2 powders (table 5.3), which have been used in sol–gel processing of colloids (chapter 4). These fumed materials are manufactured by reacting $SiCl_4$ vapour in an inert gas carrier in a H_2/O_2 (or air) stationary flame (Liu & Kleinschmit, 1986) as indicated in the overall equation

$$SiCl_4 + 2H_2 + O_2 \rightarrow SiO_2 + 4HCl \qquad (5.11)$$

and have the appearance of fluffy powders. The flame-hydrolysis route to oxide powders is limited by the availability of volatile starting materials.

5.4 Outline of the sol–gel process for alkoxides

The susceptibility of alkoxides to hydrolysis, which can complicate their synthesis, forms the basis of the sol–gel process for metal–organic compounds. This route is outlined in figure 5.1 and may be compared with the sol–gel process for colloids (figure 4.5). Partial (mole ratio of $H_2O/OR < z$) or total ($H_2O/OR \geq z$) hydrolysis of an alkoxide, referred to as the sol, liberates alcohol and results in gel formation either as a stiff monolith, fibres, coatings or powders. It is common practice to dilute the alkoxide with an alcohol before hydrolysis and, unlike sol–gel transitions involving colloids, the sol–gel conversion for alkoxides is irreversible. A useful definition of metal–organic compounds given by Yoldas (1977) refers to molecules in which organic groups are bound to a metal atom via oxygen. Hence formates, $M(CHOO)_z$, acetates, $M(CH_3COO)_z$ and acetylacetonates, $M(CH_3COCHCOCH_3)_z$ are also in this category. Some alkoxides particularly those of Si, Al, Ti, B and Zr are commercially available. These are not expensive materials thus tetraethyl silicate, TEOS, $Si(OC_2H_5)_4$ (also called tetraethyl orthosilicate or tetraethoxysilane), may be purchased from chemical distributors for about £12 per kilogram, but a non-standard alkoxide, for example, $Ba(OC_2H_5)_2$, is more expensive, about £2 per gram, at the time of writing.

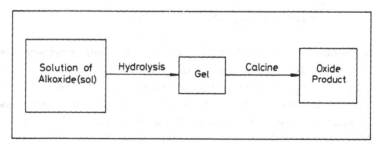

Figure 5.1. Outline of the sol–gel process for alkoxides.

Development of the sol–gel process for alkoxides

In order to prepare multicomponent oxide ceramics, alkoxides are blended together in a solvent at the sol stage and doped with metal salts if the corresponding alkoxide is unavailable, although *in situ* syntheses can sometimes be performed before hydrolysis. Gel calcination with loss of water, alcohol and decomposition of added salts results in the oxide product.

5.5   Development of the sol–gel process for alkoxides

Sol–gel processing of alkoxides is closely associated with non-melt routes to glasses and other oxide ceramics. The synthesis of oxide powders by using alkoxide hydrolysis was pioneered by Roy (1955, 1956) during a study of phase equilibria in mineral systems, for example the MgO–SiO_2–Al_2O_3 system, under hydrothermal conditions up to 1198 K. Several synthetic methods were considered, dry and wet grinding oxide components, repeated fusions combined with grinding, coprecipitation of hydroxides and ignition of mixed nitrate crystals. Evaporation of solvent from ethanolic solutions of TEOS containing aluminium and magnesium nitrates produced a rigid gel, which was converted to oxide at 673 K after decomposition of salts. The alkoxide route produced mixtures more homogeneous than could be obtained by other methods, was applicable to a wide range of mineral compositions and was easy to carry out.

Many techniques have been used to investigate the sol–gel and gel–oxide transitions particularly for TEOS. Tetraethylsilicate is immiscible with water ·and a mutual solvent, ethyl alcohol, is used to promote miscibility. Monolithic gels derived from TEOS have the same volume as the alkoxide reagents and are called alcogels in contrast to the term hydrogel or aquagel used to describe gels obtained fom aqueous sodium silicate solutions (Klein, 1985). The general approach used in these investigations has been to vary H_2O/OR ratios, the amount of acid catalyst (HCl), base catalyst (NH_4OH), solvent content and reagent concentrations. Techniques include high-resolution ^{29}Si nuclear magnetic resonance (NMR) spectroscopy at atmospheric pressure (Pouxviel *et al.*, 1987), at pressures up to 500 MPa for hydrolysis of $Si(OCH_3)_4$ (Artaki *et al.*, 1985) and magic-angle spinning (MAS) NMR spectroscopy on gels (Klemperer, Mainz & Millar, 1986). Small-angle X-ray scattering (SAXS) (Brinker *et al.*, 1984; Keefer, 1984), light-scattering (Tewari *et al.*, 1986), gas chromatography (Blum & Ryan, 1986), size-exclusion liquid chromatography (Yoldas 1986), specific and intrinsic viscosities

(Sakka & Kamiya, 1982), Raman spectroscopy and nitrogen-adsorption isotherms (Brinker *et al.*, 1985), X-ray diffraction and infra-red spectroscopy (Congshen *et al.*, 1984) and transmission electron microscopy (Mulder *et al.*, 1986) have all been used to elucidate the chemical processes involved in these transitions.

Base-catalysed reactions involve nucleophilic substitution, whereas hydrolysis in acidic solutions occurs by an electrophilic reaction mechanism (figure 5.2; Keefer, 1984). Nucleophilic attack is sensitive to both the electron density around the central Si atom and steric effects due to the size of substituent groups; susceptibility to nucleophilic attack increases with a decrease in bulky and basic alkoxy groups around the central Si atom. However, reactivity of the tetrahedron towards electrophilic attack is enhanced by an increase in electron density around silicon. Initial hydrolysis of silicon ester monomer, $Si(OR)_4$, produces silanol groups ($\equiv Si-OH$) whereas full hydrolysis can in

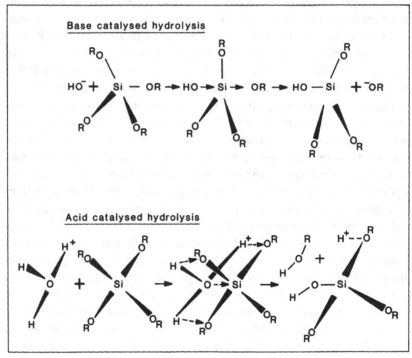

Figure 5.2. Schematic representation of reaction mechanisms for acid and base catalysed hydrolysis of alkyl silicates (Keefer, 1984).

principle lead to silicic acid monomer Si(OH)$_4$ (figure 5.3). This acid is not stable and condensation of silanol groups occurs leading to polymer formation before all alkoxy groups are substituted by silanol groups. Condensation–polymerisation reactions proceed with an increase in viscosity of the alkoxide solution until an alcogel is produced. In general, acid-catalysed reactions yield alcogels whereas base hydrolysis precipitates hydrated SiO$_2$ powders. A recent review on the hydrolysis of silicon esters has been made by Klein (1985), and Iler (1979) has described the aqueous chemistry of silicate ions including polymerisation processes.

Alumina hydrates derived from alkoxides constitute another one-component system that has been characterised (Yoldas, 1973, 1975a, b). Aluminium *sec*-butoxide and aluminium isopropoxide were hydrolysed at 293 K and 353 K with H$_2$O/alkoxide mole ratios of about 200. The higher temperature always produced fibrils of boehmite *ca*. 100 nm in length, as determined by transmission electron microscopy. However, the cold-water reaction led to an amorphous alumina monohydrate. The latter contained alkoxy groups, and crystallised to bayerite, β-Al$_2$O$_3$.3H$_2$O by a dissolution–recrystallisation process or to boehmite, above 353 K, although temperature affected the rate of hydrolysis of the alkoxides; boehmite powders could be peptised with HCl to stable sols with Cl/Al ratios 0.038–0.246 (Yoldas, 1975b). The schematic reaction scheme for hydrolysis of an aluminium alkoxide leading to a polymer containing *f* Al ions and *g* OR groups is shown in figure 5.4. Aluminium alkoxides manufactured by oxidation of aluminium alkyls (equation (5.10)) are hydrolysed to pseudoboehmite powders and marketed in Europe under the names of Dispural and Pural (Condea Chemie GmBH). When peptised, these hydrates have been used for preparing

$$Si(OC_2H_5)_4 + 4H_2O \longrightarrow Si(OH)_4 + 4C_2H_5OH$$

$$Si(OH)_4 \longrightarrow SiO_2 + 2H_2O$$

Figure 5.3. Schematic representation of hydrolysis of tetraethoxysilane.

alumina-based abrasive grain (chapter 4). This example illustrates how metal–organic compounds can be used in sol–gel processing of colloids.

Cross-condensation reactions between different metal alkoxides resulting in M–O–M' bonds at ambient temperature are illustrated in figure 5.5, and for silicon esters partial hydrolysis is carried out before adding other alkoxides because of their lower reaction rates. Interest in this sol–gel process gained momentum since the work of Dislich (1971), which showed that gels derived from mixed solutions of magnesium methoxide and aluminium *sec*-butoxide gave broad X-ray diffraction peaks for spinel ($MgAl_2O_4$) after calcining at 523 K, with a crystallite size of 10 nm from line broadening at 893 K. By comparison, spinel formation for reaction of $Mg(NO_3)_2.6H_2O$ and $NH_4Al(SO_4)_2.12H_2O$ started at 1123 K and between 623 and 673 K for coprecipitated hydroxides. In addition, borosilicate glass with composition 86.9 weight % SiO_2; 5.9 weight % B_2O_3; 2.6 weight % Al_2O_3; 3.9 weight % Na_2O; 0.7 weight % K_2O was prepared as a rigid gel from $Si(OCH_3)_4$, $Al(O^sC_4H_9)_3$, $NaOCH_3$, KOC_2H_5 and $B(OC_2H_5)_3$. When dried at 423 K the gel disintegrated to crumbs that sintered to hard clear glass pieces at 803 K. The crumbs could be hot-pressed at 903 K under 280 MPa to a monolithic plate identical in structure to fused glass.

Low-temperature sintering of alkoxide-derived glasses can result in carbon contamination because of incomplete decomposition of alcohols

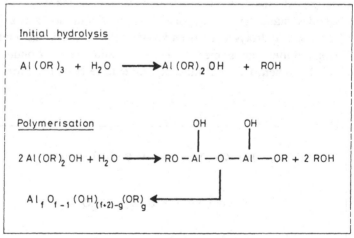

Figure 5.4. Schematic representation of hydrolysis of an aluminium alkoxide (Yoldas, 1973).

Development of the sol–gel process for alkoxides

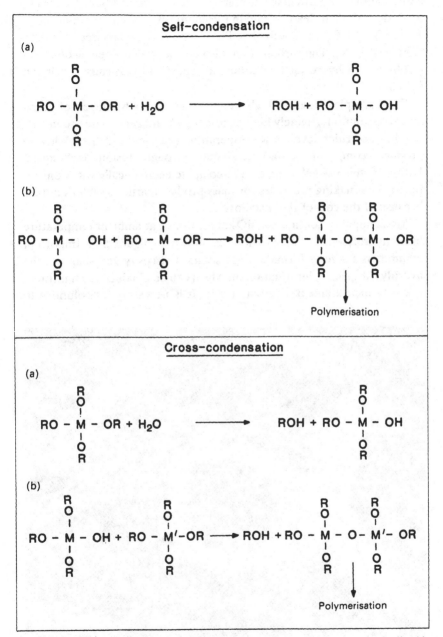

Figure 5.5. Schematic representation of condensation reactions in mixed alkoxide solutions.

67

and added salts, particularly acetates. Substitution of PbO powder (1–5 μm diameter) for lead acetate in the composition 13.5 weight % SiO_2; 13.5 weight % B_2O_3; 73 weight % PbO led to a clear glass free of carbon (Thomas, 1974). The carbon could also be eliminated when alcohol was distilled from hydrolysed solutions, an operation that caused gelation (Thomas, 1977).

The examples described above illustrate the advantages of using alkoxide solutions, namely high purity, high homogeneity due to mixing on the molecular level, low-temperature sintering and the ability to produce compositions not available through fusion techniques. However, this sol–gel route cannot compete economically with conventional glass-making processes for mass-produced articles such as bottles because of the cost of the reactants.

An example of continuous SiO_2 fibre, drawn at ambient temperature from hydrolysed TEOS solution, is shown in figure 5.6. Important parameters for fibre formation are solution viscosity and shape of the hydrolysed species, information on which can be obtained from intrinsic viscosity measurements (Sakka, 1985). It is necessary for solutions to

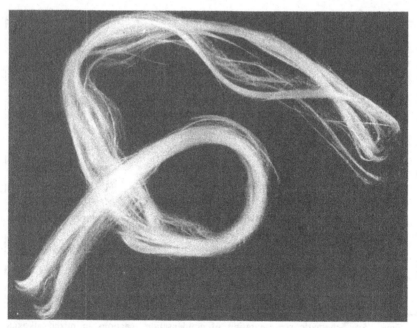

Figure 5.6. Silica fibres drawn from hydrolysed tetraethoxysilane (Sakka, 1985).

have a viscosity of more than 1 Pa s and contain linear polymers, which eliminates base-catalysed reactions as these produce particles. Gel fibres sinter directly to glass when calcined at about 1073 K in air.

There has been considerable interest throughout the worldwide research effort on sol–gel processing in producing monolithic alcogels that yield, after suitable heat treatment, dense, crack-free bodies for applications including optical components such as lenses; this conversion eliminates the need to sinter powders. However, alcogels and monolithic gels derived from colloids (chapter 4) have a tendency to crack and crumble when left to dry at ambient temperatures unless special precautions are taken, for example, controlling solvent vapour pressure by slow evaporation of solvent through pinholes in a bottle (3.8 cm diameter, 4.0 cm high) containing an alcogel derived from methanolic $Si(OCH_3)_4$ solutions (Yamane, Aso & Sakaino, 1978). Gelation occurred in about a day and dry crack-free gel one-eighth of the initial gel volume was produced after two weeks. These gels had nitrogen surface areas greater than 300 m^2 g^{-1}, volume porosities of about 30 volume % and pore diameters between 1.5 and 10 nm as determined from gas-adsorption data.

A pressure difference Δp exists across the liquid–air interface for menisci in the pores of a drying gel and is given by the Laplace equation (Defay *et al.*, 1966)

$$\Delta p = 2\gamma \cos\theta / r_p \qquad (5.12)$$

where r_p is the pore radius, γ the surface tension of liquid and θ the contact angle at the solid–liquid–air boundary. Variation of this capillary pressure with pore radius is shown for H_2O in figure 5.7 assuming perfect wetting ($\cos \theta = 1$). A consequence of the capillary pressure is a stress acting on walls of pores in the gel whose magnitude increases with decreasing pore size and which can lead to collapse of capillaries and crack formation in the dry gel.

Shoup & Wein (1977) showed that monolithic gels could be produced from alkali-metal silicate solutions when an organic additive such as formamide, $HCONH_2$, was added before gelation. As an example, potassium silicate solution containing $HCONH_2$ gelled when mixed with a quaternary ammonium silicate. The hydrogel was immersed in NH_4NO_3 to reduce pH and leach out K^+ ions after which 80% of pores had diameters within 30% of the mean diameter, 90 nm, determined by mercury porosimetry. The total pore volume was 74 volume % and transparent monoliths were obtained on sintering in He at 1723 K; oxides made by heating hydrogels are also called xerogels when completely dehydrated. The studies on aqueous systems have been extended by Hench (1986) together with co-workers (Orcel & Hench, 1984; Wallace & Hench, 1984) for addition of drying control chemical additives (DCCAs) such as formamide and oxalic acid to alkoxide solutions. The effect of DCCAs on the pore-size distribution of silica gels is shown in figure 5.8, and figure 5.9 illustrates crack-free monoliths made by this method. Detailed mathematical descriptions of drying processes in gels have been given by Scherer (1987).

Stresses responsible for cracking of gels disappear if liquid menisci in pores are eliminated and this can be achieved by using supercritical or

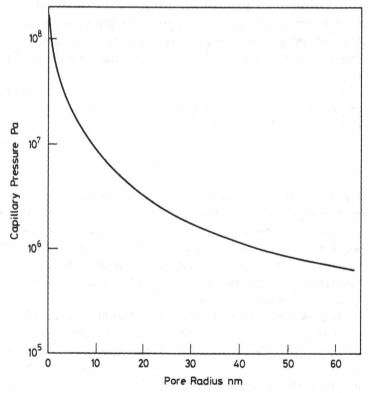

Figure 5.7. Variation of capillary pressure with pore radius for a water-filled capillary (Zarzycki, 1984).

hypercritical drying conditions in an autoclave. A supercritical fluid is a substance above its critical temperature where it remains as a single fluid phase and although increasing pressure affects fluid density it will not produce a separate liquid phase. Examples of critical temperatures and pressures for water and selected organic solvents are shown in table 4.7; solids made by supercritical drying of alcogels and aquagels are known as aerogels. The method was devised by Kistler (1932) whereby water was displaced from a hydrogel by alcohol before placing the alcogel in an autoclave and removing vapour under supercritical conditions. It is necessary to substitute alcohol for water because inorganic oxides are soluble in aqueous media at the temperatures and pressures used for drying, but preparation of alcogels directly from alkoxides avoids this solvent exchange process. Supercritical drying has been used (Teichner, 1986) for preparing single oxide (SiO_2, Al_2O_3) and mixed oxide (Al_2O_3; Cr_2O_3, TiO_2; MgO) powders with high surface area and porosity for applications including catalyst supports and rocket propellants; silica aerogels can have surface areas of 1000 $m^2 g^{-1}$, pore volumes of 20 $cm^3 g^{-1}$ and particle diameters *ca.* 2 nm.

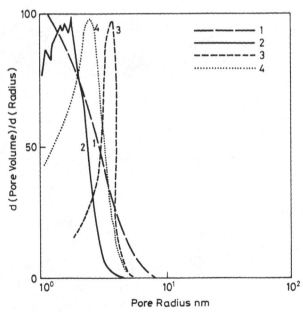

Figure 5.8. Pore-size distribution of dried silica gel monoliths (Hench, 1986).
(1) Tetramethoxysilane/water/methyl alcohol, (2) HCl catalysed, (3) 50 volume % formamide DCCA and (4) oxalic acid DCCA.

Sol–gel processing of metal–organic compounds

The ability to produce monolithic shapes (Zarzycki, 1984; van Lierop *et al.*, 1986) depends on the heating rate, amount of solvent, alkoxide concentration, the H_2O/alkoxide ratio, geometry and size of the article together with ageing of the gel. Supercritical drying produces a porous monolith with identical volume to the alcogel, avoiding shrinkage, which occurs on drying at ambient temperature; examples of aerogel composition are 15 mole % B_2O_3; 85 mole % SiO_2 and 20 mole % B_2O_3; 5 mole % P_2O_5; 75 mole % SiO_2 (Woignier, Phalippou & Zarzycki, 1984). Monoliths densify and convert to glasses on further heat treatment and examples are shown in figure 5.10.

The use of aqueous sols for preparing crystalline ceramics as alternative to glass for solidification of high-level radioactive liquid waste was described in section 4.8. Waste is dissolved in a melt at over 1273 K for conventional glasses, but the melting point limits the composition and high temperatures can result in loss of volatile radioactive components. Glasses have been developed (Pope & Harrison, 1981) by mixing either all metal alkoxide reactants, or Si and B alkoxides together with additives in the form of hydroxides. Waste simulate was added to the

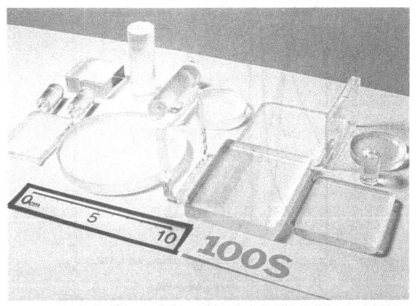

Figure 5.9. Formamide modified silica gel monoliths dried at 333 K (Wallace & Hench, 1984).

mixed alkoxide solutions and gel obtained on drying was fused to a glass at 1273–1473 K. In this way, alumina-containing glasses (*ca.* 28 weight % Al_2O_3) with high durability, that would be difficult to make by fusion of powders because of high forming temperatures (1623 K), were prepared. The glass-ceramic composition 6.6 mole % Na_2O; 5.1 mole % Al_2O_3; 16.5 mole % CaO; 14.8 mole % TiO_2; 57.0 mole % SiO_2 was made (Vance, 1986) with $Ti(OC_2H_5)_4$, colloidal SiO_2 sol, sodium nitrate, aluminium nitrate and calcium nitrate. Gel obtained on drying reactants was ball-milled in isopropanol after calcination at 873 K to decompose nitrates and organics. Whereas conventional fabrication of this glass ceramic involves melting oxides at *ca.* 1673 K followed by crystallisation at 1323 K, crystals of sphene, $CaTiSiO_5$ (also called titanite) about 1 μm diameter formed in the ball-milled powder at 1103 K. The sol–gel sphene ceramic, which has a high durability in granitic groundwaters, was pressed under about 50 MPa and sintered at 1173 K to *ca.* 97% of theoretical density (2780 kg m^{-3}) with less than 1% open porosity. This example illustrates how aqueous sols and alkoxide solutions can be used in the same reaction mixture for synthesising ceramics.

Sol–gel processing of metal–organic compounds has been applied to the synthesis of non-oxide ceramics, and in an early study (Hoch & Nair,

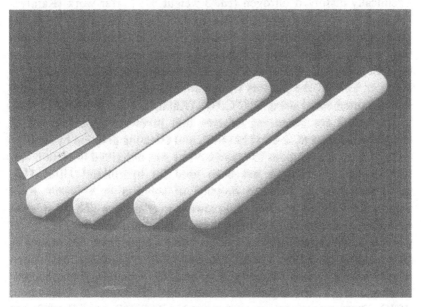

Figure 5.10. Aerogel obtained by supercritical drying (van Lierop *et al.*, 1986).

Sol–gel processing of metal–organic compounds

1979) sub-micrometre α-Si$_3$N$_4$ was obtained after nitriding silica gel, derived from TEOS, in dry NH$_3$(g) at 1623 K. However, sol–gel SiO$_2$ produced an amorphous material when heated in NH$_3$(g) at 1273 K that crystallised to α-Si$_3$N$_4$ in N$_2$ at 1623 K, but cristobalite formed when heated directly in N$_2$ at the latter temperature. Hydrolysis of Si(OC$_3$H$_7$)$_4$ containing carbon black (size 0.07 μm) produced a uniform distribution of C in SiO$_2$ that had a particle size 0.02 μm determined from transmission electron microscopy (Mitomo & Yoshioka, 1987). Sub-micrometre (0.9 μm) α-Si$_3$N$_4$ containing 0.8 weight % carbon and 1.2 weight % oxygen was obtained after nitriding gel in N$_2$ at 1723 K; a similar route was used for 100% AlN with particle size 0.6 μm and carbon content 1.6 weight % starting from Al(OC$_2$H$_5$)$_3$. Conventional routes to oxynitride glasses involve exposing melts and porous borosilicate glasses to anhydrous ammonia together with dissolution of additives such as Si$_3$N$_4$ in melts (Loehman, 1985). In an attempt to prepare oxynitride ceramics (Brinker, 1982) a monolithic gel (pore volume *ca.* 0.1 cm^3 g^{-1}, surface area 300–500 m^2 g^{-1}, pore size 3–4 nm) with representative composition 69.9 weight % SiO$_2$; 18.0 weight % B$_2$O$_3$; 6.9 weight % Al$_2$O$_3$; 1.6 weight % Na$_2$O; 3.6 weight % BaO was exposed to NH$_3$(g) at 773 K, which raised the sintering temperature from 848 K to 878 K because of chemically dissolved nitrogen (*ca.* 3 weight %). Later work (Kamiya, Ohya & Yoko, 1986) showed that up to 6 weight % N$_2$ can be incorporated into SiO$_2$ fibres when the gel is heated in NH$_3$(g) at 1073 K and that nitrided fibres have a higher chemical resistance than silica in alkaline solution. Titanium nitride fibres have also been prepared by firing, at 1273 K, the product from ammonolysis of TiO$_2$ gel fibres (100 μm diameter) made using Ti(O^{i}C$_3$H$_7$)$_4$ (Kamiya, Yoko & Bessho, 1987).

A method for preparing spheres that involved dispersing sol to droplets in a solvent, 2-ethyl hexanol, and effecting gelation by transfer of H$_2$O from the aqueous to organic phase was described in section 4.6. An analogous technique has been used for hydrolysed Ti(OC$_2$H$_5$)$_4$ (H$_2$O/alkoxide = 2) whereby droplets of solution in kerosene were rapidly dehydrated by using a microwave oven after which gel spheres could be separated from the organic phase (Komarneni & Roy, 1985). A potential application for these oxide spheres, diameter *ca.* 30 μm, was as plasma-spray powders. Oxides for electronic applications have been synthesised by using alkoxide solutions, for example Nasicon compounds, which are superionic conductors (Perthuis & Columban, 1986). Representative compositions of Na$_3$Zr$_2$Si$_2$PO$_{12}$ and Zr-deficient

$Na_{3.1}Zr_{1.55}Si_{2.3}P_{0.7}O_{11}$ were prepared either as transparent monoliths or powders 0.2–0.5 μm diameter after firing gels at 1273 K. Further examples of electronic ceramics in the form of coatings are described in the next section.

5.6   Alkoxide-derived coatings

Reflectivity is the ratio of reflected to incident light intensity for inter-action at a boundary separating two media with refractive indices n_0, n_1 and its value R^* for near-normal incidence is given by the expression

$$R^* = [(n_1 - n_0)/(n_1 + n_0)]^2 \qquad (5.13)$$

A coating with thickness h_c and refractive index n_2, which separates media with indices n_0, n_1, modifies the magnitude of R^* so that

$$R^* = [(n_2^2 - n_1 n_0)/(n_2^2 + n_1 n_0)]^2 \qquad (5.14)$$

For minimum reflectivity at wavelength λ

$$n_2 = (n_1 n_0)^{1/2}, \text{ and} \qquad (5.15)$$

$$h_c = \lambda/4n_2 \qquad (5.16)$$

which results in destructive interference for light reflected from front and back faces of the film. Hence coatings, either single or multiple, can modify the reflectivity of surfaces and hydrolysed alkoxide solutions have been used widely for depositing films by techniques that include spinning, dipping, spraying and lowering (whereby substrate remains stationary and the liquid is lowered). On exposure to damp air the films convert to gel layers, which are calcined to oxide coatings. Alkoxide solutions allow control of refractive index in the coating as shown in figure 5.11. In addition, low surface tensions encourage wetting of sub-strates while adhesion to glass is promoted by the hydroxylated nature of the surface.

Development of single oxide films by dipping techniques was pioneered by Geffcken & Berger (1939) although vacuum deposition was a competitive process at that time. However, renewed interest in alkoxide-derived coatings has arisen since the work of Schroeder (1962, 1969) and is sustained at the present time. That work was on single oxides, In_2O_3, ThO_2 (using nitrate solutions), Al_2O_3, TiO_2 (using alkoxides) and mixed oxides SiO_2–TiO_2.

Laser-induced damage to antireflectance coatings on optical components such as lenses and mirrors by high-powered pulsed lasers can result in pit formation, which limits power output. The damage threshold of multilayer coatings formed by vapour deposition is 5 J cm^{-2} for 1 ns pulses at 1.06 μm but higher damage thresholds, 20 J cm^{-2}, have been obtained (Mukherjee, 1981; Mukherjee & Lowdermilk, 1982) by using sol–gel coatings. In this work, a gel layer 32 nm thick with a corresponding oxide composition 84 weight % SiO$_2$; 12 weight % B$_2$O$_3$; 4 weight % Na$_2$O was prepared by dip-coating with partially hydrolysed Si(OCH$_3$)$_4$ containing B(OCH$_3$)$_3$, NaOCH$_3$ and allowing liquid to gel in air. Multiple layers were built up and calcined at 623–773 K to an oxide with microporous structure. This heat treatment produced a leachable phase that could be removed with acidified NH$_4$F to give a graded refractive index in the coating. Apart from enhanced damage threshold an advantage of the graded-refractive-index coatings compared with multiple films was their antireflectance properties over a wide wavelength range, 0.35–2.5 μm. Silicate–phosphate glasses have also been used for protective laser filters (Dislich & Hinz, 1982).

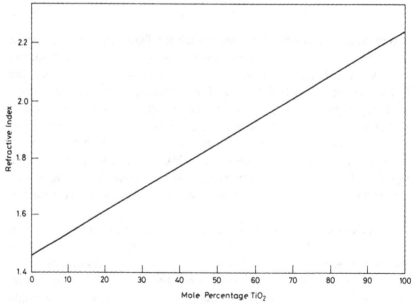

Figure 5.11. Variation of refractive index at 550 nm with SiO$_2$–TiO$_2$ composition as a function of TiO$_2$ content (Dislich, 1985a).

Alkoxide-derived coatings

Antireflectance films have been deposited on Si wafers to increase the efficiency of solar cells as up to 35% of the incident radiation can be reflected from the bare substrate (Yoldas & O'Keeffe, 1979). Film thickness depends on rate of application and solution properties, particularly viscosity, which is affected by concentration, degree of hydrolysis, ageing, liquid medium and by added polymer for example hydroxypropylcellulose (Zelinski & Uhlmann, 1984). The composition (95–30) mole % TiO_2; (5–70) mole % SiO_2 was applied by spinning a mixed $Ti(OC_2H_5)_4/Si(OC_2H_5)_4$ solution which, after firing at temperatures below 723 K, resulted in amorphous oxide films less than 100 nm thick with refractive indices between 1.63–2.17 (Brinker & Harrington, 1981), whereas dip-coating has been used for a 88% TiO_2; 12% SiO_2 composition (Yoldas & O'Keefe, 1979). The variation of reflectivity with wavelength for a coating thickness of about 75 nm in the latter system is shown in figure 5.12. Kern & Tracy (1980) sprayed $Ti(O^iC_3H_7)_4$ (7.7 volume %) in a mixture of *n*-butylacetate (17 volume %), 2-ethyl hexanol (33 volume %) and *sec*-butanol (42 volume %) onto Si whereas the lowering technique, which is suited for large substrates, has been used (Ashley & Reed, 1984) on glass solar receiver envelopes, 3 m long

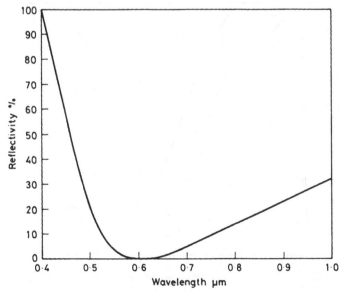

Figure 5.12. Variation of reflectivity with wavelength for an antireflectance coating on a silicon solar cell (Yoldas & O'Keeffe, 1979).

and *ca.* 6 cm diameter. A porous oxide with composition 71 weight % SiO_2; 18 weight % B_2O_3; 7 weight % Al_2O_3; 4 weight % BaO obtained on firing gel at 773 K was acid-etched to produce a single layer coating whose reflectivity was 0.03–0.05 compared with 0.09 for untreated tubes with a corresponding minimum value at 550 nm.

The deposition of coatings on window glass has received much attention and these films have to transmit both visible light and solar energy but reflect long-wavelength (9.5 μm) infra-red (IR) radiation (Schroeder 1969; Dislich, 1985a). Solar transmittance and IR reflectivity of single-layer semiconductor films are related by the free-carrier concentration, film thickness and carrier mobility (Arfsten, Kaufmann & Dislich, 1984). An increase in conductivity enhances reflectivity and shifts the plasma frequency to shorter wavelength with a corresponding decrease in solar transmittance because of increased adsorption in the film; free charge carriers and lattice defects affect adsorption in the visible region. High mobility together with conductivity produce the desired IR reflectivity and optical properties for a conducting indium tin oxide, ITO, (6 mole % In) dip-coat, as shown in figure 5.13. Cadmium stannate is another electrically conducting coating that can be obtained by dip

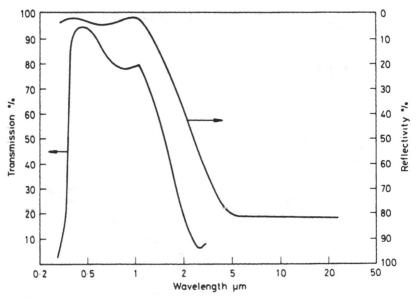

Figure 5.13. Reflectivity and transmission for an indium tin oxide coated glass (Arfsten *et al.*, 1984).

techniques with mixed solutions of $Cd(CH_3COO)_2$ and $Sn(O^nC_4H_9)_4$. However, oxide deposited by sputtering in vacuum or spraying is not single-phase material in contrast with sol-gel Cd_2SnO_4 (Dislich & Hinz, 1982). Glass substrates coated with ITO are manufactured and marketed by the Schott Glasswerke (Federal Republic of Germany). A different system that has been used for window glass is b K_2O; $(1-b)$ Na_2O; Al_2O_3; $4SiO_2$; $2/3$ B_2O_3, where $b = 0$, 0.2, 0.4, 0.6, 0.8 or 1 (Strawbridge, Phalippou & James, 1984) and film thicknesses of about 300 nm were obtained by dip-coating.

Alkoxide-derived coatings have not been used exclusively for anti-reflectance layers. Coloured films with particular optical absorbance properties have been deposited onto glass by dipping so that transition metal oxides are incorporated into the substrate. Hence after baking at 653–723 K, films from ethanolic solutions of $Fe(OC_2H_5)_3$ were amber-red in colour, those from cobalt acetate solutions were dark grey–amber and coatings from solutions of chromium acetate in ethyl alchol were light grey–green (Geotti-Bianchini *et al.*, 1984). The metal (Cr, Mn, Fe, Co, Ni and Cu) can be introduced as the nitrate salt to TEOS prior to deposition of silica coatings so that film thicknesses 0.3–0.5 μm result after firing at *ca.* 773 K (Sakka *et al.*, 1983). Recent work on titania-based ceramics has shown that adherent yellow coatings about 600 nm thick and containing a crystalline phase could be prepared on glass and Al foil (Makishima *et al.*, 1986) after firing the gel film, derived from Ce(III) chloride dissolved in $Ti(O^iC_3H_7)_4$ (Ce/Ti mole ratio 1) at 773 K. Glassy layers of GeO_2–SiO_2 obtained from $Si(OCH_3)_4$/$Ge(OC_2H_5)_4$ and containing up to 30 weight % GeO_2 have increased the oxidation resistance of porous silicon carbide at temperatures up to 1473 K (Schlichting & Neumann, 1982). Pores were impregnated with alkoxides that were then hydrolysed to coatings and similar results have been obtained for porous molybdenum disulphide layers on niobium substrates by using this infiltration technique.

Fibres have also been coated with alkoxides. The reinforcement of ordinary Portland cement with glass fibres can be affected by lack of chemical resistance of the fibres in alkaline media. This resistance can be enhanced by incorporating ZrO_2 into the fibres but is an expensive process. Guglielmi & Maddelena (1985) have shown that a sol–gel 20 weight % ZrO_2; 80 weight % SiO_2 coating on the fibre deposited by dipping can produce the same chemical resistance as conventional alkali-resistant fibres. Sol-gel coatings have not been restricted to oxide sub-

Sol–gel processing of metal–organic compounds

strates as SiC fibres and whiskers were coated with alumina (Lannutti & Clark, 1984).

The ability to produce chemically pure oxides makes the sol–gel method particularly attractive for preparing thin films of interest to the electronics industry. Hence tantalum oxide on Si wafers by spinning partially hydrolysed solutions of $Ta(OC_2H_5)_5$ in ethyl alcohol (H_2O/Ta mole ratio 1–1.3) for use as dielectrics in capacitors (Silverman, Teowee & Uhlmann, 1986), magnetic iron oxide dip coats on borosilicate glass (Kordas, Weeks & Arfsten, 1985), SnO_2–iron oxide and yttrium iron oxide films 50–155 nm thick on glass (Bruce & Kordas, 1986). Electrically conducting Sb_2O_3-doped In_2O_3 (8 weight % Sb_2O_3 or less) films for use as transparent electrodes have been prepared (Puyane & Kato, 1983) by dipping after which the gel was fired up to 903 K, while $BaTiO_3$ layers 100–300 nm thick were deposited on Si, Ni, Ti substrates (5 cm diameter) by spinning a mixture of $Ti(O^iC_3H_7)_4$ and methanolic $Ba(OH)_2$ solution (Dosch, 1984); the latter method allowed a greater control of stoichiometry than sputtering techniques. The mixture used for depositing $PbZr_{0.5}Ti_{0.5}O_3$ and $PbLa_{1-b}(Zr_{0.65}Ti_{0.35})_{1-b/4}O_3$ ($b = 0.09$) films less than 1 μm thick illustrates an *in situ* synthesis of alkoxides (Budd, Dey & Payne, 1985). Solutions of Pb and La acetate hydrates in methoxyethanol, $CH_3OCH_2CH_2OH$, were dehydrated at the solvent boiling point, 398 K. Alkoxides were added after which solvent and isopropylacetate were distilled off leaving a mixed complex alkoxide that was applied, when partially hydrolysed, by spinning onto Si and fused SiO_2 substrates.

Non-oxide films have been much less investigated than oxides although Si_3N_4 and oxynitride films are attracting increasing attention because the latter on Si enhance oxidation resistance of this substrate more than SiO_2 films and can be used as an oxidation barrier during microelectronic device fabrication. Two approaches have been used, incorporation of Si_3N_4 into an alkoxide solution and nitridation of oxide films. In the first method (Martinsen, Figat & Shafer, 1984), α-Si_3N_4 powder, 0.5 μm diameter, was dispersed in an alkoxide, which was spun onto both fused quartz and sapphire substrates. Films 7–10 μm thick with a representative composition 65 weight % Si_3N_4; 25 weight % Al_2O_3; 10 weight % Si were obtained on firing up to 1573 K. The second method (Haaland & Brinker, 1984; Brow & Pantano 1986) involved nitridation of gel films in dry ammonia; for example oxynitride coatings 50 nm thick were obtained on firing the gel, deposited by spinning TEOS onto Si

80

wafers, at 1073–1373 K. Oxidation resistance for these films increased with firing temperature because of enhanced uptake of nitrogen whose content after 1273 K was 40 mole %.

When a liquid coating derived from a metal–organic compound, for example, TEOS, dries to a gel there is an approximately 50 weight % loss from the time of application until gelation occurs and an additional 50 weight % loss on drying the gel; these combined weight losses correspond to a volume reduction of *ca.* 70%. Also, shrinkage occurs on drying in the thin dimension, not in the plane of the substrate (Klein, 1984). However, coatings are prone to crack on drying the gel and conversion to oxide with loss of adhesion (figure 5.14). The causes of cracking include thermal expansion mismatch between coating and substrate, stresses due to shrinkage on firing and phase transformations although under certain conditions crack-free layers can be produced, for example by applying multiple gel films before calcination (Budd *et al.*, 1985).

Figure 5.14. Example of a cracked silica coating derived from hydrolysed tetraethoxysilane on a glass substrate (Strawbridge, 1984).

5.7   *Monodispersed sub-micrometre oxide powders*

Since 1980 increasing attention has been paid to the preparation of monodispersed sub-micrometre oxide powders with a view to obtaining denser packing than can be achieved with non-spherical material and lower sintering temperatures. However, the principles that determine formation of monodispersed particles were established nearly 40 years ago when La Mer & Dinegar (1950) considered three reaction regions to describe growth in acidified sodium thiosulphate and in ethanolic sulphur solutions diluted with water (figure 5.15). The molecular sulphur concentration, C_x, increased above the saturation value, C_s, during an initial induction period until a critical supersaturation concentration, C_{ss}, was reached in time t_1 after mixing reactants. Homogeneous self-nucleation then occurred followed by particle growth involving diffusion

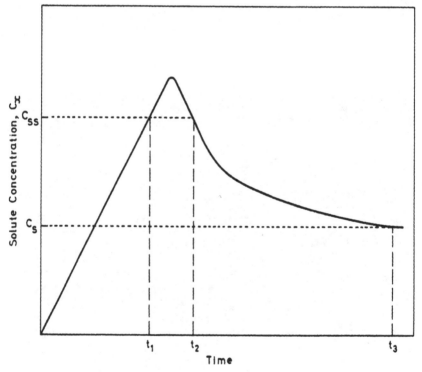

Figure 5.15. Schematic representation of solute concentration versus time before and after nucleation (LaMer & Dinegar, 1950).

of molecular S to the nuclei up until t_3 when $C_x = C_s$. Monodispersed sulphur sols were produced when one burst of nucleation occurred in a short time interval, t_2–t_1, which was achieved by using low (10^{-2} M or less) reactant concentrations. Uniform growth took place when molecular S was released slowly because diffusion to nuclei occurred without build-up of solute concentration and further bursts of nucleation. Sulphur sols with particles 0.60–0.98 μm in diameter were prepared in this way with induction and reaction times of about 1 min and 25 min, respectively. Diameters were measured by changes in angular position with particle size of red bands in the higher-order Tyndall spectra. This qualitative approach to homogeneous nucleation is applicable to sol formation by cation hydrolysis (section 4.4), hydrothermal processing of solutions, vapour-phase reactions and the controlled hydrolysis of alkoxides.

Stöber, Fink & Bohn (1968) carried out a systematic study on the hydrolysis of silicon esters and prepared monosized unaggregated spherical SiO_2 particles 0.05–2 μm diameter, measured using transmission electron microscopy, by controlled growth of silicic acid. For different solvents, the reaction rate was fastest in methanol and slowest in *n*-butanol whereas particle size increased on changing from methanol to *n*-butanol. However, for different silicates, fastest reaction times (less than 1 min) and smallest sizes (less than 0.2 μm) occurred with $Si(OCH_3)_4$ whereas the reaction time for tetrapentylsilicate was about 24 h, which resulted in a distribution of diameters around 2 μm. A representative ester concentration was 0.28 mol dm^{-3} for TEOS corresponding to 17 g dm^{-3} SiO_2, a significantly smaller oxide concentration than used in sol–gel processing of colloids (chapter 4).

Studies on the synthesis and characterisation of monosized unaggregated spherical powders involving controlled hydrolysis of metal–organic compounds have continued on single and mixed oxides. Barringer & Bowen (1982) prepared amorphous hydrated TiO_2 with particle sizes between 0.07–0.3 μm and 0.3–0.6 μm, determined from scanning electron microscopy, using $Ti(O^iC_3H_7)_4$ and $Ti(OC_2H_5)_4$ respectively; spheres were only obtained with ethoxide, isopropoxide yielding equiaxed particles. Specific surface areas of TiO_2 powders were sensitive to the washing procedure (Barringer & Bowen, 1985a). Water-washed material gave areas 250–320 m^2 g^{-1}, much higher than alcohol-washed oxide because a layer of unreacted alkoxide hydrolysed, resulting in a surface coating. Spherical particles were obtained only when four conditions were satisfied. First, reagents were filtered for removal of insoluble impurities to prevent heterogeneous nucleation, then mixed thoroughly at H_2O/alkoxide ratios of 3 or more before particle nucleation. Thirdly, concentrations of alkoxide, 0.1–0.2 M and water, 0.3–1.5 M in alcoholic solutions were used to produce a single homogeneous nucleation process and finally rapid growth of nuclei to particles 100 nm diameter was necessary for avoiding aggregation by van der Waals forces. Oxide spheres made by controlled hydrolysis are shown in figure 5.16.

Doped-TiO_2 powders have been made by co-hydrolysis of mixed alkoxides and by precipitation of hydroxide or carbonate in a TiO_2 sol prepared from sub-micrometre powders; dopant was added to sol either as an alkoxide or metal salt solution (Fegley & Barringer, 1984; Barringer *et al.*, 1984). In this way Ba, Cu, Nb, Sr and Ta were intro-duced into TiO_2 for concentrations up to 1 weight %.

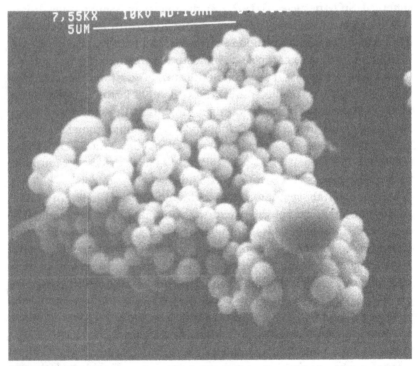

Figure 5.16. Titania spheres by controlled hydrolysis of an alkoxide. (Courtesy of the UKAEA.)

Monodispersed sub-micrometre oxide powders

An insight into the mechanism for hydrolysis of $Ti(OC_2H_5)_4$ was gained (Barringer & Bowen, 1985a) from variation of the time for onset of turbidity with H_2O concentration assuming that the rate-controlling hydrolysis species accumulated until nucleation occurred at a concentration C_{ss}. For this species

$$\text{rate of formation} \propto [H_2O]^3[Ti(OC_2H_5)_4] \qquad (5.17)$$

suggesting a hydrolysis reaction

$$Ti(OC_2H_5)_4 + 3H_2O \rightleftharpoons Ti(OC_2H_5)(OH)_3 + 3C_2H_5OH \qquad (5.18)$$

although equation (5.18) does not exclude the presence of dimers or trimers besides the monomeric hydrolysis species $Ti(OC_2H_5)(OH)_3$. Polymerisation of monomer to hydrated oxide is represented by the equation

$$Ti(OC_2H_5)(OH)_3 \rightarrow TiO_2.bH_2O + (1-b)H_2O + C_2H_5OH \qquad (5.19)$$

with an overall reaction the sum of equation (5.18) and 5.19)

$$Ti(OC_2H_5)_4 + (2+b)H_2O \leftrightarrows TiO_2.bH_2O + 4C_2H_5OH \qquad (5.20)$$

where thermogravimetric analysis indicated b was between 0.5 and 1.0.

Monosized spheres in the zirconia system have also been prepared (Ogihara, Mizutani & Kato, 1987; Tormey et al., 1984); typical parameters were 0.50 M H_2O, 0.10 M alkoxide and a reaction time of 2 min. The method was extended to 6.3 mole % Y_2O_3–ZrO_2 spheres, 100 nm diameter, using co-hydrolysis of an $Y(O^iC_3H_7)_3$ solution in $Zr(O^nC_3H_7)_4$ at 323 K (Fegley & Barringer, 1984; Fegley, White & Bowen, 1985). The rate of hydrolysis for boron alkoxides, faster than for TEOS, decreases with an increase in alkyl chain length. In order to dope SiO_2 uniformly with boron, TEOS was partially hydrolysed with NH_4OH (Barringer et al., 1984), then tri-n-butyl borate was added (B/Si ratio 1) after SiO_2 spheres had grown to 85% of their final size (ca. 250 nm). The effect of a polymeric steric stabiliser (section 4.3) hydroxypropylcellulose (HPC, molecular weight 6×10^4) on nucleation and growth from $Ti(OC_2H_5)_4$ was investigated (Jean & Ring, 1986) for HPC concentrations between 0.05 and 3.4 g dm^{-3}. Polymer did not act as a nucleating agent but stabilised particles by surface adsorption, which reduced the activation energy for homogeneous nucleation through decrease of the interfacial free energy; diameter, number

Table 5.4. *Comparison of sintering temperature between monodispersed and conventional milled powders (Barringer & Bowen, 1985b)*

Oxide	Initial size/μm	Average final grain size/μm	Sintering temperature/K	
			Monodispersed	Conventional milled
TiO_2	0.3	0.5	1273	1873
ZrO_2	0.2	0.3	1273	1973
Al_2O_3	0.25	0.5	1523	2023
SiO_2	0.4		1373	
B_2O_3–SiO_2	0.2		ca. 973	

density and agglomeration of particles were all affected by this adsorption process. Recent preparative studies on sub-micrometre SiO_2 spheres have been described by Milne (1986).

The synthesis of ZnO-based varistor materials using sol–gel processing of colloids was described in section 4.8. An alternative approach to the preparation of ZnO has been taken by Heistand & Chia (1986) who hydrolysed ethylzinc *t*-butoxide, $(C_2H_5)Zn(O^tC_4H_9)$ in C_2H_5OH/H_2O mixtures at room temperature. Water concentration and washing procedure (aqueous or alcoholic) determined particle morphology although spherical polycrystalline ZnO was obtained in a narrow size distribution with diameter 0.2 μm, crystallite size 15 nm and surface area 30 m^2 g^{-1}.

The achievement of lower sintering temperatures using sub-micrometre spherical particles with a narrow size distribution compared with conventional milled powders is shown in table 5.4 (Barringer & Bowen, 1985b). Fast precipitation times (minutes), simple solution chemistry and high yields (*ca.* 95%) indicate that controlled hydrolysis reactions are suitable for scale-up. If alkoxide concentrations are limited to *ca.* 0.2 M and cannot be significantly increased before aggregation and non-spherical powders are produced then the cost of scale-up due to using large liquid volumes may outweigh advantageous sintering properties of these powders. However, monosized powders have been made on the pilot-plant scale with yields of hundreds of grams per hour (Tormey *et al.*, 1984).

5.8 Organically modified silicates

The sol–gel process described in this chapter involves using alkoxides in which all attached groups can be replaced on hydrolysis after which gel calcination at *ca.* 700 K removes organic groups yielding a wholly inorganic compound. However, a class of materials, organically modified silicates or Ormosils (Scholze, 1985), can be prepared when some groups are not hydrolysable so that the reaction product is an organic–inorganic composite that is not subjected to a heat treatment. These materials have mechanical properties intermediate between glasses and plastics and examples of their chemical structure are shown in figure 5.17; methylpolysiloxane has been used as a coating on optical fibres (Dislich, 1985b). Another application for Ormosils has been as materials in hard contact lenses (Philipp & Schmidt, 1984; Schmidt & Philipp, 1985) that

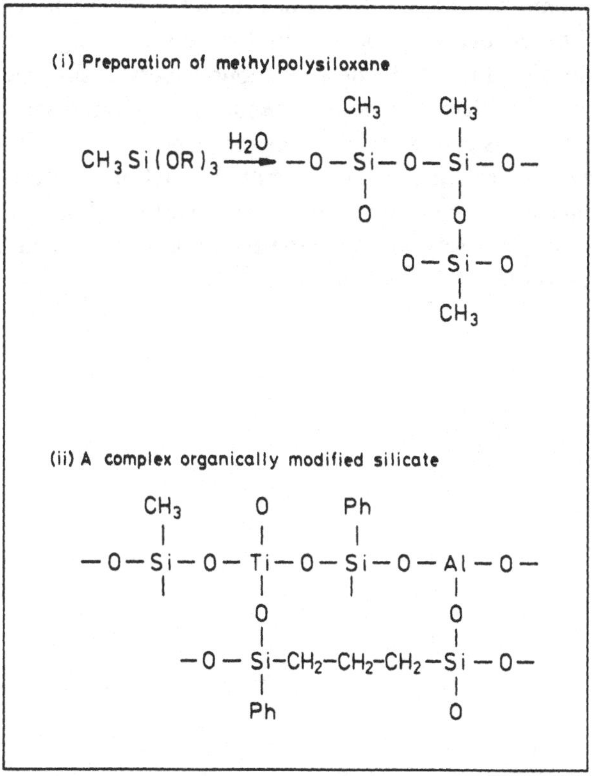

Figure 5.17. Examples of organically modified silicates (Dislich, 1985b).

have good wettability, permeability to oxygen, flexibility and scratch resistance. A related class of materials, ceramers, has been described by Huang, Orier & Wilkes (1985) and are flexible transparent hybrids made by reacting, for example, polydimethylsiloxane (molecular weight 1700) with TEOS so that cross-condensation occurs between ester and the silicone polymer.

5.9   Summary

Sol–gel processing of metal–organic compounds, principally alkoxides, has attracted immense interest during the past 15 years with hundreds of published papers on the subject, a small representative selection of which is included in this chapter. The method offers a low temperature non-melt route to high-purity glasses and crystalline ceramics. Certain applications, for example sol–gel coatings on window glass, have been placed on a commercial basis. Alkoxides allow the preparation of crack-free monolithic alcogels, aerogels and xerogels avoiding sintering of powders and also the production of monodispersed sub-micrometre spherical oxide particles by homogeneous nucleation from solution. Hydrolytic reactions can be carried out on reagents that contain some non-hydrolysable groups, which result in hybrid materials with properties intermediate between glasses and plastics. However, sol–gel processing of alkoxides cannot compete economically with conventional fusion routes to mass produced articles.

6 Non-aqueous liquid-phase reactions

6.1 Introduction

The title of this chapter refers to synthesis of advanced ceramic powders by using non-aqueous liquid media, which can be inert or constitute a reactant, for example ammonia, and at the time of writing is particularly associated with silicon nitride. There has been increasing interest during the 1980s in developing low-temperature liquid-phase syntheses, which has resulted in the industrial production of certain powders. Reactions between ammonia and silicon tetrachloride are described in this chapter, together with those involving other chlorosilanes, ammonia and amines. Finally, the preparation of non-oxide powders, as well as Si_3N_4, with liquid-phase reactions are outlined. Work described here and in chapter 9 highlights the importance of controlling reaction conditions in order to reduce impurity levels such as chlorine in Si_3N_4. The significance of these levels may not be appreciated when synthesis is viewed strictly from a chemical approach, and without also considering the ceramic applications of, for example, high-temperature structural components.

6.2 The reaction between silicon tetrachloride and ammonia in the liquid phase

Persoz (1830) carried out initial studies on the liquid-phase reaction between silicon tetrachloride and ammonia. He considered that the white precipitate obtained on passing $NH_3(g)$ through $SiCl_4$ dissolved in benzene at *ca.* 273 K was silicon tetramide, $Si(NH_2)_4$, a conclusion also drawn by Lengfeld (1899) and later by Vigoureux & Hugot (1903). However, this precipitate was unstable and lost $NH_3(g)$ at ambient temperature to give silicon diimide, $Si(NH)_2$, a polymeric species that was not separated from ammonium chloride, and similar products were

Non-aqueous liquid-phase reactions

obtained by Glemser & Naumann (1959) who carried out the reaction in liquid ammonia.

Initial products are complex and although the reaction can be written as

$$SiCl_4(l) + 6NH_3(g) \rightarrow Si(NH)_2 + 4NH_4Cl \qquad (6.1)$$

work by Billy (1959) indicated a more involved sequence

$$SiCl_4 + 18NH_3(g) \rightarrow 1/f(Si(NH)_2)_f + 4NH_4Cl.3NH_3 \qquad (6.2)$$

resulting in polymeric silicon diimide and ammonium chloride triammoniate, $NH_4Cl.3NH_3$. Silicon diimide decomposes according to the overall equation

$$3Si(NH)_2 \rightarrow Si_3N_4 + N_2 + 3H_2 \qquad (6.3)$$

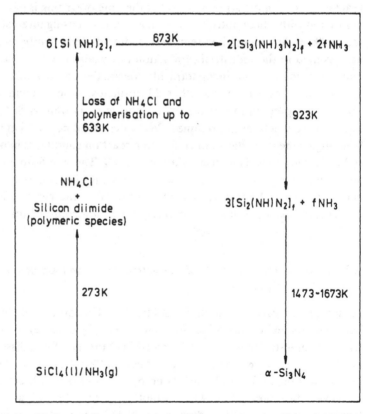

Figure 6. 1. Possible intermediates formed during the preparation of α-silicon nitride from silicon tetrachloride and ammonia (Segal, 1986).

90

Reaction between silicon tetrachloride and ammonia

Investigations by Glemser & Naumann (1959), together with those by Mazdiyasni & Cooke (1973, 1976) who reacted $SiCl_4(l)$ and $NH_3(g)$ in dry hexane at 273 K, indicated that polymeric intermediates may have been formed when initial reaction products underwent vacuum decomposition as shown in figure 6.1. In contrast with these observations, Billy and his co-workers (1975) claimed that halide products such as $NH_4Cl.3NH_3$ affected the thermal stability of Si (and Ge) imides. Silicon diimide was stable up to 473 K in a vacuum or inert atmosphere (N_2 or NH_3) when halides were extracted with $NH_3(l)$, but continual loss of $NH_3(g)$ occurred at higher temperatures until Si_3N_4 formed at about 873 K. However, decomposition yielded imidohalides in two stages when halides were present. Condensation occurred initially at 333–353 K, for example,

$$2Si(NH)_2 \rightarrow Si_2N_3H + NH_3 \qquad (6.4)$$

followed by reaction between the intermediate, Si_2N_3H, and partially dissociated NH_4Cl vapour at 403–423 K as indicated in the overall equation

$$2Si(NH)_2 + NH_4Cl \rightarrow Si_2N_3H_2Cl + 2NH_3 \qquad (6.5)$$

and the imidohalide, $Si_2N_3H_2Cl$ yielded Si_3N_4 above 1073 K.

Mazdiyasni & Cooke (1973, 1976) showed, using X-ray measurements, that α-Si_3N_4 could be prepared from the initial reaction products as a fine powder (particle diameter 10–30 nm). The latter was amorphous to X-rays for decomposition up to 1473 K but transformed to the α-phase on prolonged heating between 1473 and 1673 K, for example 8 h at 1473 K and to β-Si_3N_4 at 1723 K. Very pure nitride could be prepared with total impurities less than 300 p.p.m., while Fe and Ni, the principal contaminants, were each present at less than 100 p.p.m. Amorphous Si_3N_4 powder with a needle-like morphology was uniaxially vacuum hot-pressed under 41 MPa at 1923 K in the presence of a 4 weight % Mg_3N_2 sintering aid to β-Si_3N_4, which had a fine grain, uniform microstructure and 98.5% of theoretical density (3200 kg m^{-3}).

The reaction between $SiCl_4(l)$ and $NH_3(l)$ has been scaled-up by Ube Industries (Iwai, Kawahito & Yamada, 1980) and their process is shown schematically in figure 6.2. $SiCl_4(l)$ was introduced below the surface of an organic solvent (e.g. 80 weight % cyclohexane; 20 weight % benzene) with a density greater than ammonia, which constituted the second liquid phase. An interfacial reaction occurred at 233 K on agitating the liquids and $NH_4Cl.3NH_3$ was extracted into $NH_3(l)$. Reaction products were collected by filtration, washed with $NH_3(l)$ and calcined initially at 1273 K to an amorphous powder and then at 1823 K in N_2 to α-Si_3N_4. This manufacturing route does not generate NH_4Cl fumes and also avoids blockages in inlet pipes as $SiCl_4$ was discharged directly into the solvent. Properties of an industrial Si_3N_4 powder made by this synthesis are listed in table 6.1, and a scanning electron micrograph for the same material (UBE-SN-E10 grade) is shown in figure 6.3.

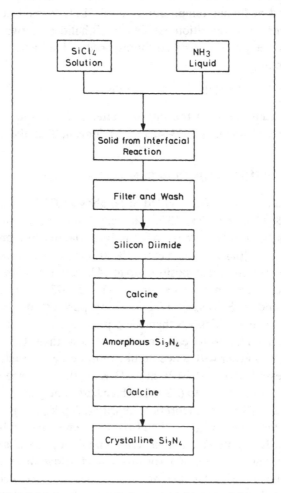

Figure 6.2. Schematic representation of a commercial process for producing silicon nitride powder by a liquid-phase reaction (Kohtoku *et al.*, 1986).

Table 6.1. *Typical properties of a commercially available silicon nitride powder made by a liquid phase reaction (Kohtoku et al., 1986; Yamada, Kawahito & Iwai, 1984)*

Grade	UBE-SN-E10
Morphology	Equiaxed
Surface area/($m^2 g^{-1}$)	10
Crystallinity	100%
α-Si_3N_4 (weight %)	>95
β-Si_3N_4 (weight %)	<5
Metallic impurities (weight %)	Fe (0.005); Ca (<0.001); Al (0.002)
Non-metallic impurities (weight %)	O (1.3); C (<0.1); Cl (0.005)

Comparison of table 6.1 with table 9.3 shows that lower impurity levels can be obtained in Si_3N_4 powders derived from liquid-phase reactions than from other syntheses.

The morphology of α-Si_3N_4 powders made in liquid-phase reactions is, in general, controlled by calcination times and temperature (Franz,

Figure 6.3. Scanning electron micrograph for a commercially available silicon nitride powder (UBE-SN-E10 grade) made by a liquid-phase reaction (Kohtoku *et al.*, 1986).

Schönfelder & Wickel, 1986). Fine-grained particles with shapes similar to silicon diimide can be obtained at lower temperature, while needle-like and coarse-grained hexagonal particles form above 1773 K.

Reaction between $SiCl_4(l)$ and $NH_3(l)$ is a highly exothermic process. Although thermodynamic data are not available, standard heat and free energy of reactions (at 298 K) for the related but hypothetical reaction

$$SiCl_4(g) + 6NH_3(g) \rightarrow Si(NH)_2 + 4NH_4Cl(s) \qquad (6.6)$$

have been calculated by Crosbie (1986) as -676 kJ mol^{-1} and -384 kJ mol^{-1}. Heat from the reaction can be dissipated by using organic solvents, for example, hexane (Mazdiyasni & Cooke, 1973). However, an alternative approach has been adopted by Crosbie (1986, 1987) and researchers at the Ford Motor Company that avoided solvents and hence the possibility of carbon contamination on calcination. Nitrogen gas at 0.5 MPa brought $SiCl_4$ vapour into contact with $NH_3(l)$ at 273 K and this pressurised system has two advantages. First, refrigeration costs are reduced at these higher operating temperatures and secondly the solubility of $NH_4Cl.3NH_3$ in liquid ammonia increases with temperature (figure 6.4). In addition, heat from the reaction was approximately

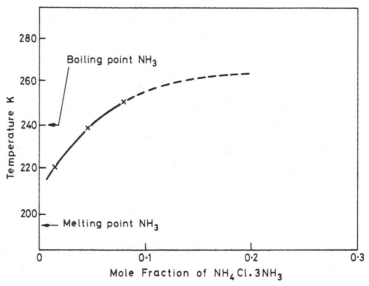

Figure 6.4. Solubility of ammonium chloride triammoniate in liquid ammonia (Billy *et al.*, 1970).

balanced by the latent heat of vaporisation for liquid ammonia, a useful property for future economic scale-up of the process. Equiaxed Si_3N_4 powder with an α-phase content between 85 and 95 weight %, particle diameter 0.2–0.3 μm, nitrogen surface area 23 m^2 g^{-1} and containing 0.08 weight % C, 0.06 weight % Al, 0.16 weight % Fe, 0.06 weight % Ca and 0.03 weight % Ti was synthesised in this process (Crosbie, 1987).

6.3 General reactions of chlorosilanes with ammonia and amines

Besides reacting with NH_3, chlorosilanes yield silylamines with primary or secondary amines (Eaborn, 1960) for example,

$$(CH_3)_3SiCl + 2NH(C_2H_5)_2 \rightarrow (CH_3)_3SiN(C_2H_5)_2 \\ + (C_2H_5)_2NH_2Cl \qquad (6.7)$$

or silazanes, compounds containing Si–N–Si bonds such as hexamethyl-disilazane (Eaborn, 1960) and hexaphenylcyclotrisilazane (Larsson & Bjellerup, 1953; Mazdiyasni, West & David, 1978). The stability of silylamines with respect to silazane formation increases with functional group size and structures of representative silazanes are shown in figure 6.5.

Silicon diimide is a non-volatile insoluble solid and although Si_3N_4 powders are manufactured from it there is growing interest in preparing liquid silazanes that convert with high yield (and minimal shrinkage) on pyrolysis to monolithic shapes. This approach is analogous to work described in section 5.5 whereby crack-free monoliths were obtained

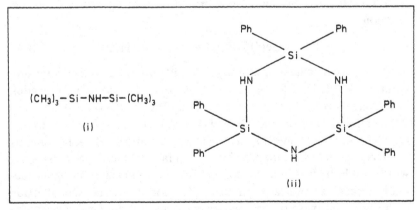

Figure 6.5. Structure of representative silazanes (i) hexamethyldisilazane and (ii) hexaphenylcyclotrisilazane.

directly from alcogels by using slow evaporation of solvent, drying control chemical additives and supercritical drying. Stock & Somieski (1921) showed that dichlorosilane, H_2SiCl_2, reacted with NH_3 in benzene solution forming a viscous oil after ammonium halide products were removed. Cryoscopic molecular-weight measurements indicated a value of approximately 350 and the unstable oil set to a hard clear solid after 24 h. The reaction between H_2SiCl_2 and NH_3 occurs more readily in polar solvents such as dichloromethane or diethyl ether (Seyferth, Wiseman & Prud'homme, 1983, 1984) yielding the polysilazane oil $(H_2SiNH)_j$ as observed in earlier studies (Stock & Semieski, 1921). Si–N bonds in this polymer are sensitive to hydrolysis but the silazane appeared to be indefinitely stable under N_2 at 243 K although its viscosity increased at room temperature over several days when it set to a glassy solid. The oil could be pyrolysed with 70% yield at 1423 K in N_2 to α-Si_3N_4 pieces *ca.* 0.2 cm in length with density 2500 kg m^{-3} and grain size *ca.* 0.2 μm. Polymer pyrolysis routes to ceramic materials are described further in the next chapter.

6.4 Synthesis of non-oxide ceramics other than silicon nitride using liquid-phase reactions

Liquid-phase reactions that allow low-temperature synthesis of ceramic precursors (e.g. silicon diimide) have been extended to other materials besides Si_3N_4. In a novel application of sol–gel processing (Melling, 1984), amorphous germanium sulphide was prepared by reacting $H_2S(g)$ with $Ge(OC_2H_5)_4$ solution at room temperature according to the reaction

$$2H_2S + Ge(OC_2H_5)_4 \rightarrow GeS_2 + 4C_2H_5OH \qquad (6.8)$$

and crystallised on heating although sulphide was not obtained after substituting $Ge(SC_2H_5)_4$ for germanium ethoxide. The postulated reaction mechanism involved nucleophilic attack by HS^- on the central Ge atom. Polarisation is less in the Ge–S bond than in Ge–O, which lowers reactivity for the former towards nucleophilic attack. Back bonding by 3d electrons from Ge into 3d orbitals on S favours the Ge–S interaction, which would be lost in a five-coordinate transition state during nucleophilic attack, as indicated in figure 5.2 for base-catalysed alkoxide hydrolysis, and these two properties could account for lack of reactivity in $Ge(SC_2H_5)_4$ (Melling, 1984). Low impurity levels that can be obtained with sol–gel materials (table 5.2) are desirable for this synthesis as GeS_2 has application in infra-red transmitting optical fibres, while this liquid-phase reaction occurs at lower temperature than the conventional synthesis using vacuum heating of Ge and S powders.

An aggregated cubic β-ZnS powder for infra-red transmitting ceramics was obtained (Johnson, Hickey & Harris, 1986) with primary particles 100 nm or smaller in size on passing $H_2S(g)$ through a diethyl zinc solution in heptane at room temperature as indicated by the equation

$$Zn(C_2H_5)_2 + H_2S \rightarrow ZnS + 2C_2H_6 \qquad (6.9)$$

Powder was vacuum heated at 1073 K after compaction under 800 MPa, but resulting pellets contained carbon residues from the solvent and the ethyl groups. These small ZnS particles (100 nm or less in dimension) can lead to ceramics with grain sizes less than the infra-red wavelength (8–12 μm) eliminating light scattering between grains. A feature common both to this reaction and sol–gel processing of metal–organic compounds (chapter 5) is that components are mixed on the molecular level.

Reductive dechlorination of halide solutions has also been used (Ritter, 1986) to synthesise precursors for boride and carbide powders. For example, $SiCl_4$ and CCl_4 reacted with Na in heptane at 403 K as shown in the equation

$$SiCl_4 + CCl_4 + 8Na \rightarrow \text{'SiC' precursor} + 8NaCl \qquad (6.10)$$

to give NaCl and a black amorphous precursor that crystallised between 1723 and 2023 K in 5% H_2/Ar to SiC with particle size 1–5 μm and surface area 26 $m^2 \, g^{-1}$. A possible mechanism involved initial generation of nucleophilic species (e.g. Cl_3Si^-), which formed a complex with another halide molecule so that the product contained mixed-metal bonds and this process continued until all Cl was eliminated leaving a matrix that constituted the precursor. This synthetic route was also applicable to TiB_2 and B_4C, but for boron nitride NH_4Cl was reacted with sodium borohydride in benzene between 443 and 453 K (Kalyoncu, 1985). Although not isolated, the product was considered to be a borane or borazine and yielded crystalline BN on calcination in N_2 at 1373 K. The aggregated powder (figure 6.6) had a surface area between 80 and 100 $m^2 \, g^{-1}$ and a corresponding primary particle size 20–30 nm consistent with dimensions obtained from X-ray line broadening. Techniques used for manufacture of Si_3N_4 have been applied to other metallic nitrides

(Iwai *et al.*, 1980). Hence the product from reaction of vanadium tetra-chloride with $NH_3(l)$ in a heptane-toluene mixture yielded cubic vanadium nitride with a particle size 0.1–1 μm when decomposed in NH_3 at 1773 K.

6.5   Summary

Liquid-phase reactions allow low-temperature synthesis of precursors for non-oxide ceramics, principally carbides, sulphides, borides and nitrides, and processes have been scaled-up for manufacture of silicon nitride. High-purity, sub-micrometre isolated particles with a narrow size distribution that can be obtained in these reactions are desirable powder properties for fabrication of structural components

Figure 6.6. Scanning electron micrograph for boron nitride powder prepared in a liquid-phase reaction and calcined at 1373 K (Kalyoncu, 1985).

7 Polymer pyrolysis

7.1 Introduction

Formation of polymeric cationic species in aqueous colloidal dispersions was described in section 4.4, while polymers obtained in hydrolysed tetraethyl silicate solutions have been referred to in section 5.5. The contents of this chapter, however, are concerned with synthesis and properties of polymeric ceramic precursors derived from chlorosilanes, namely polysilanes, polycarbosilanes and polysilazanes, together with their conversion, on pyrolysis, to ceramics mainly in the form of fibres and coatings on both the laboratory and industrial scale. This work, which lies at the organic–inorganic interface, illustrates the contribution that different branches of chemistry have made to ceramic synthesis, and also how an area of organosilicon chemistry, not immediately associated with ceramics, can make a significant impact on materials preparation.

7.2 Synthesis of polysilanes

Unlike conventional organic polymers, such as polystyrene, which contain a chain of carbon atoms or silicones consisting of alternating Si and O atoms, the polymer backbone in polysilanes comprises Si–Si linkages (figure 7.1). Early work on aryl-substituted polysilanes was reported by Kipping and co-workers in a series of publications during the 1920s. As an example, diphenyldichlorosilane, $(C_6H_5)_2SiCl_2$, was reacted (Kipping & Sands, 1921) with molten sodium in an inert solvent, xylene, and several products were obtained. These included a sparingly soluble solid, $Si_4(C_6H_5)_8$, which they called an unsaturated compound and the major product, a soluble material with the same formula and referred to as a saturated compound; solvent evaporation left a resinous mass from the latter. These formulas were derived from elemental analysis, but low solubilities for the products prevented molecular-weight determination. In the case of alkyl-substituted polysilanes, Burkhard (1949) showed that dimethyldichlorosilane, $(CH_3)_2SiCl_2$ would react with molten Na in benzene at 473 K under 1.5 MPa according to the equation

$$(CH_3)_2SiCl_2 \xrightarrow{\text{Na}} NaCl + CH_3((CH_3)_2Si)_gCH_3 \qquad (7.1)$$

A white crystalline polydimethylsilane, $CH_3((CH_3)_2Si)_gCH_3$, was the major product while a minor soluble fraction remained in benzene solution. Molecular-weight determination with biphenyl as solvent indicated that the degree of polymerisation (g in equation (7.1)) was 55 for this permethylated polysilane and six for the soluble fraction,

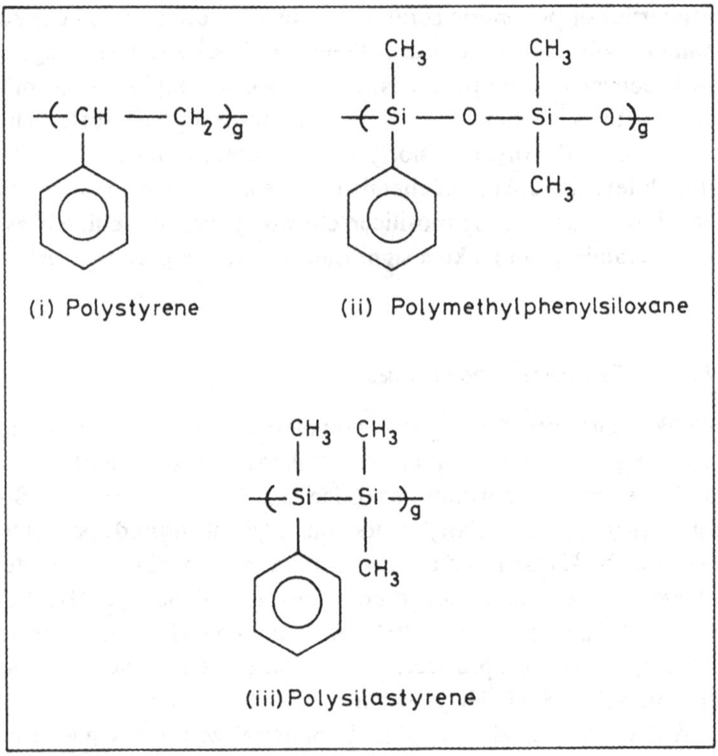

Figure 7.1. Structures of typical polymers (i) an organic polymer, (ii) a silicone polymer and (iii) polysilastyrene (West, 1984)

100

corresponding to a six-membered cyclic oligomer, dodecamethylcyclo-hexasilane, $((CH_3)_2Si)_6$. However, interest in the solid polydimethyl-silane as a ceramic precursor did not exist at that time because the polymer had very limited solubility and was thermally unstable, con-verting to a gelatinous material when heated in air at 473 K for 16 h.

Polysilanes have strong absorption bands in the wavelength region 300–350 nm characteristic of transitions among delocalised σ-electrons in Si–Si bonds. However, the mechanism for polymerisation, whether occurring at the metal surface or in solution has not been fully elucidated although initial reaction may involve electron transfer from Na to produce a chlorosilane anion or radical (West, 1986).

7.3 The polysilane–polycarbosilane conversion

The pioneering work of Yajima and associates showed how β-SiC fibres could be prepared from polysilanes. Dimethyldichlorosilane was con-verted to polysilanes by two routes. $((CH_3)_2Si)_6$ was synthesised in high yield by reacting lithium chips with $(CH_3)_2SiCl_2$ in tetrahydrofuran (Yajima *et al.*, 1976a; Yajima, Hayashi & Omori, 1975a) and purified by recrystallisation and sublimation, whereas the second route involved pre-paration of polydimethylsilane according to equation (7.1) using xylene as solvent (Yajima *et al.*, 1976b). Polysilanes were converted to poly-carbosilanes, polymers containing repeating Si–C units, by a suitable heat-treatment. The Li-derived material underwent ring-opening and polymerisation at 673 K under Ar in an autoclave, whereas poly-dimethylsilane was also converted at elevated pressure and at atmos-pheric pressure under Ar at 593 K to a liquid that was refluxed for several hours at 743 K yielding the highly viscous polycarbosilane; solid polymer was obtained after dissolving the latter in hexane followed by solvent evaporation. Polycarbosilanes can be prepared from polysilanes besides those derived from $(CH_3)_2SiCl_2$. Octaphenylcyclotetrasilane and hexamethyldisilane precursors have been described in the patent litera-ture (Yajima, Hayashi & Omori, 1978a), while Fritz & Matern (1986) have given a detailed account of polycarbosilane synthesis.

The Si–Si bond in polysilanes will cleave first because its energy, 222 kJ mol^{-1}, is less than the value of 318 kJ mol^{-1} for Si–C, 314 kJ mol^{-1} for Si–H and 414 kJ mol^{-1} for C–H, so that the mechanism for formation of the carbosilane (polysilapropylene) shown in figure 7.2 involves cleavage of Si–Si followed by radical rearrangement (Yajima *et al.*, 1978b). Also,

Polymer pyrolysis

Si–C bonds are more resistant to cleavage in free-radical reactions than C–H bonds and polymer growth can occur by dehydrogenation–condensation processes. Hence, heat treatment of polydimethylsilane at 733 K yielded polycarbosilane with an empirical formula $SiC_{1.83}O_{0.03}H_{4.6}$ compared with SiC_2H_6 for polysilapropylene indicating a lower H content in the former polymer. Information on the compositon and structure of polycarbosilanes has been obtained from their infra-red (IR) and NMR solution spectra (Yajima *et al.*, 1978b); figure 7.3 illustrates the type of ring system that can be developed in these polymers.

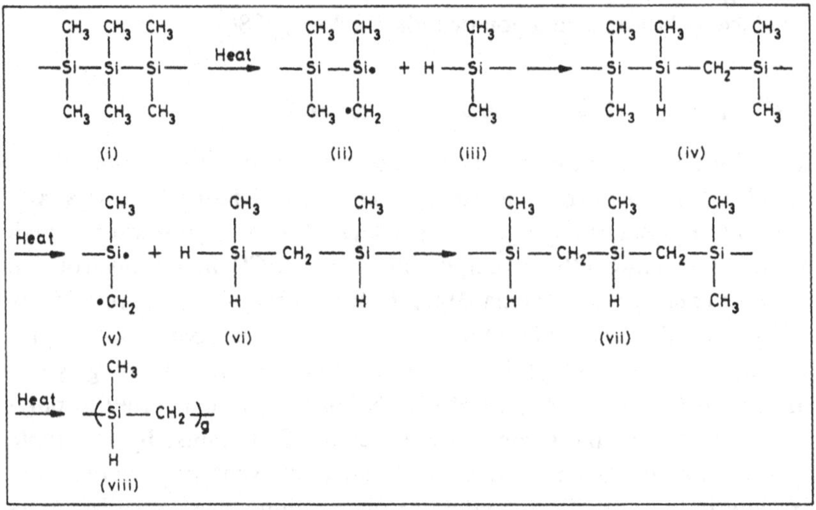

Figure 7.2. Possible mechanism for the formation of polycarbosilane from polydimethylsilane (Yajima *et al.*, 1978b).

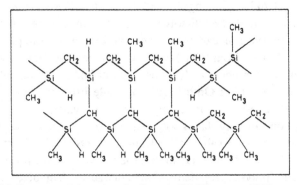

Figure 7.3. Structure of the polycarbosilane A-470 (Okamura, 1987).

102

Homogeneous nucleation processes described in sections 5.7, 8.2 and 9.2 are characterised by formation of monodispersed ceramic particles. However, polymer growth in polycarbosilanes leads to a distribution of molecular weights, and hence polymer sizes, which have been measured by gel-permeation chromatography (Yajima *et al.*, 1978b) and solvent fractionation (Yajima *et al.*, 1976c) as illustrated for the polymer obtained on heating polydimethylsilane at 723 K. Powdered poly-carbosilane was dissolved in hexane and the solution divided into soluble and insoluble fractions after adding acetone. The insoluble fraction was dissolved in hexane and fractionation repeated several times. Solids from evaporated hexane solutions were dissolved in benzene and vapour-pressure osmometry used to determine number-average molecular weights ($\bar{M}_n$). The magnitude of $\bar{M}_n$ for three fractions was 1086, 1226 and 1199 with a lower cut-off at 240, and 90 weight % of polycarbosilane had a molecular weight less than 25 000, while 52 weight % was less than 10 000. Molecular-weight distributions in poly-carbosilanes are important parameters that affect fibre formation as described in the next section.

7.4   β-Silicon carbide fibres from polycarbosilanes

Continuous polycarbosilane fibre can be prepared either by drawing from solutions or by melt-spinning, which allow control of fibre diameter. For example, filaments 10–20 μm diameter were drawn from a viscous benzene solution of polycarbosilane ($\bar{M}_n = 1500$) derived from $((CH_3)_2Si)_6$ (Yajima *et al.*, 1975a) although the fraction with $\bar{M}_n$ equal to 1199 described in section 7.1 had poor fibre-forming properties compared with the other two fractions because of a tail in its molecular-weight distribution around 25 000. Melt-spinning is not restricted to polycarbosilanes (Yajima *et al.*, 1978a) and has been applied to B_2O_3 fibres (section 9.4). However, fibre destruction occurred during subsequent vacuum heating if the polycarbosilane contained species with molecular weight between 200 and 800, which reduced the softening point of the polymer below 323 K. This was because low-molecular-weight compounds melted and aggregated with larger molecules but these small molecules could be removed from polycarbosilanes either by distillation, solvent fractionation using acetone or by polymerisation between 323 and 973 K.

Infra-red spectroscopy has been used to monitor structural changes during vacuum decomposition of fibres drawn from a polycarbosilane ($\bar{M}_n = 1500$) dissolved in benzene and synthesised from dodecamethyl cyclohexasilane (Yajima, Okamura & Hayashi, 1975b); the experimental scheme is shown in figure 7.4. Structural change began at 573 K following a decrease in the absorption band at 2100 cm^{-1} corresponding to Si–H for which cleavage was complete by 723 K producing crosslinking in the polycarbosilane. Absorption bands between 1050 and 700 cm^{-1} showed that Si–C and Si–O bonds were present at 1023 K although there was no evidence for the presence of either Si–CH_2 or Si–CH_2–Si linkages. A 40% weight loss occurred up to 1073 K when decomposition was complete and the residue only contained Si–C bonds. Randomly orientated β-SiC crystallites were formed gradually between 1073 and

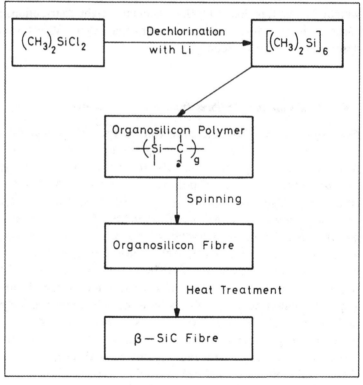

Figure 7.4. Schematic diagram illustrating the synthesis of continuous β-silicon carbide fibre (Yajima *et al.*, 1976d).

Table 7.1. *Mechanical properties of Nicalon silicon carbide fibre manufactured by polymer pyrolysis (Teranishi et al., 1983; Ishikawa & Teranishi, 1981)*

Property	Nicalon
Crystallinity	Polycrystalline
Crystal phase	β-SiC
Crystallite size (nm)	3.3 at 1573 K
Density (kg m^{-3})	2550
Filament diameter (μm)	10–15
Maximum usable temperature (K)	1523
Tensile strength (GPa)	2.45–2.94
Young's modulus (GPa)	176–196

1773 K whereby the Si–C skeleton in the fibre converted to the carbide structure during this heat-treatment.

Nicalon, a β-SiC ceramic, is manufactured by Nippon Chemical Company Limited in the form of continuous fibre and fabricated into multifilament yarns. The production route has been reviewed by Andersson & Warren (1984) and representative properties of this material are shown in table 7.1. Polycarbosilane ($\bar{M}_n = 1500$) made by dechlorination of $(CH_3)_2SiCl_2$ with sodium is melt-spun under N_2 at around 623 K and the fibres are cured at 463 K, a process that cross-links molecular chains of the polymer by O_2 and prevents melting during decomposition. Curing also causes significant oxygen uptake (*ca.* 15 weight %) into the fibre and polymer pyrolysis at 1573 K in H_2 or vacuum yields a microcrystalline ceramic.

A comparison of table 7.1 with table 4.6 shows that β-SiC has a higher tensile strength than polycrystalline alumina or vitreous aluminosilicate fibre. This non-oxide fibrous ceramic is one of the strongest materials known and an important application is in reinforced ceramic or metallic matrices (Okamura, 1987) at elevated temperature where high strength and fracture toughness are required together with both low thermal expansion and density. Reinforcement in these composites can also be achieved by using whiskers that have been synthesised by carbothermic reduction of rice hulls (Lee & Cutler, 1975) containing between 15 and 20 weight % SiO_2 and cellulose, a carbon source. The reaction proceeds at 1873 K in an Ar/CO atmosphere through formation of gaseous silicon monoxide (equations (9.8) and (9.9)) in a rate-determining step and is catalysed by iron. This temperature, considerably lower than that required in the conventional Acheson process (section 9.5), reflects intimate mixing of components in the hulls and a comparison can be made with low-temperature reactivity for sol–gel materials (chapters 4 and 5). About 10 weight % of carbide was present as whiskers *ca.* 0.3 µm diameter with lengths up to 30 µm whereas the remainder was a powder with particle size around 0.1 µm.

Polytitanocarbosilane, PTC, is a ceramic precursor related to poly-carbosilane (Okamura, 1987; Yajima *et al.*, 1981). In a representative synthesis, polyborodiphenylsiloxane, PB, was prepared by initially reacting boric acid with $(C_6H_5)_2SiCl_2$ in *n*-butyl ether under reduced pressure at 573 K (Yajima, Hayashi & Okamura, 1977) according to the equation

$$3((C_6H_5)_2SiCl_2)+2H_3BO_3 \rightarrow (C_6H_5)_6Si_3B_2O_6+6HCl \qquad (7.2)$$

after which the intermediate product $(C_6H_5)_6Si_3B_2O_6$ was pyrolysed at 573 K. This heat-treatment yielded PB whose infra-red spectrum indicated that $((C_6H_5)_2SiOBO_2)$ was a structural unit and polymerisation occurred in N_2 at 623 K between PB and polydimethylsilane obtained from dechlorination of $(CH_3)_2SiCl_2$. The resulting polycarbosilane ($\bar{M}_n = 950$) was cross-linked with $Ti(O\,^tC_4H_9)_4$ initially in xylene at 403 K and then, after solvent was distilled off, at 493 K to produce PTC ($\bar{M}_n = 1674$), which contained (Si–O–Ti) units. Boron-containing siloxane acts as a catalyst in this route to PTC, which had an empirical formula $SiTi_{0.15}C_{4.11}H_{8.80}O_{0.40}$, melted at 473 K and was soluble in organic solvents, for example tetrahydrofuran. A 27.7 weight % loss on pyrolysis occurred at 1973 K compared with 45.5 weight % for poly-carbosilane and the resulting mixed β-SiC/TiC ceramic contained an atomic ratio $SiTi_{0.15}C_{0.80}H_{0.05}O_{0.17}$.

The continuous fibre Tyranno is manufactured by Ube Industries in a process that involves pyrolysis of melt-spun PTC ($\bar{M}_n = 1300$–1400) in N_2 between 1073 and 1773 K. The mechanism for loss of high-temperature strength is complex. However, in the case of β-SiC it is associated with crystallite growth for the carbide phase and elimination of carbon and oxygen from the fibre. Above 1673 K, decrease in strength of Tyranno is caused by grain growth for β-SiC and TiC although this Ti-containing fibre appears to exhibit superior mechanical properties as its tensile strength in air diminishes above 1473 K compared with a temperature of 1373 K for β-SiC (Okamura, 1987).

7.5 Polysilastyrene

Synthesis of β-SiC from polysilanes such as $((CH_3)_2Si)_6$ requires an intermediate processing step that involves thermal polymerisation and fractionation of polycarbosilane (sections 7.3 and 7.4). West (1981), together with co-workers (West *et al.*, 1983), discovered a group of soluble phenylmethylpolysilanes, referred to as polysilastyrene, that could be pyrolysed directly to non-oxide ceramics. An exothermic reaction occurs between dimethyldichlorosilane and phenylmethyldichlorosilane in tetrahydrofuran containing Na, K or Na/K alloy (78% K). Copolymer was precipitated from the refluxing liquid after quenching the reaction by first adding hexane followed by water. Its formation can be represented by the equation

$$(CH_3)_2SiCl_2 + C_6H_5CH_3SiCl_2 \rightarrow [((CH_3)_2Si)_b(C_6H_5CH_3Si)_j]_g \quad (7.3)$$

where the $(CH_3)_2Si/C_6H_5CH_3Si$ mole ratio, $b\!:\!j$ ranged from 3:1 to 20:1 and g was between 1 and 100. This ratio affected the solubility, tractability, melting point and decomposition temperature (West, 1981; Mazdiyasni, West & David, 1978). For example, when $(CH_3)_2Si/C_6H_5CH_3Si$ was 4:1, the copolymer melted between 573 and 603 K, had the consistency of gum and was slightly soluble in toluene although a crystalline, insoluble solid that softened at high temperature was obtained for ratios greater than 5:1; copolymers were characterised by IR, ^1H and ^{13}C NMR spectroscopy. However, ratios between 0.5:1 and 1.5:1 yielded soluble amorphous solids called polysilastyrene because their structures resembled polystyrene, as shown in figure 7.1. Copolymer, dissolved in tetrahydrofuran, was fractionated by initially adding CH_3OH to precipitate phenylmethylpolysilane leaving cyclic oligomers in solution (West *et al.*, 1983). Molecular-weight fractions were then extracted from a copolymer solution with isopropanol and their composition determined by integration of phenyl and methyl proton signals in NMR spectra. Intrinsic-viscosity measurements and gel-permeation chromatography generated molecular-weight distributions, often trimodal (David, 1987) containing typical $\bar{M}_n$ values at 300 000 and 12 000 as well as oligomers of molecular weight less than 1000.

Polymer pyrolysis

Polysilastyrene can be molded, cast into films from solution and drawn to fibres. Rigid shapes for this copolymer were obtained by exposure to ultra-violet radiation, which induced cross-linking in contrast with thermal curing required for polycarbosilanes. However, an 80% weight loss that took place on pyrolysis compared with 28% for polytitanocarbosilane (Yajima *et al.*, 1977) is undesirable for maintaining physical shape during decomposition and indicates that depolymerisation was occurring in preference to cross-linking. β-SiC whiskers made by pyrolysis of a phenylmethylpolysilane in Ar at 1673 K are shown in figure 7.5.

7.6   Nitride and oxynitride fibres by polymer pyrolysis

Pyrolysis of organosilicon polymers is not limited to ceramics composed entirely of silicon carbide. Hence the silylamine (section 6.3)

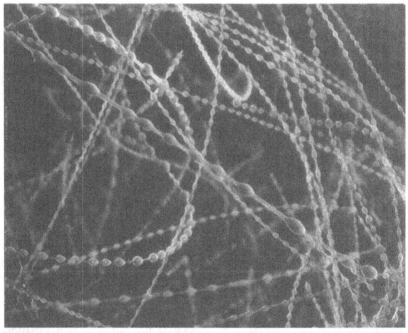

5 μm

Figure 7.5. β-Silicon carbide whiskers from pyrolysis of a phenylmethylpolysilane (Mazdiyasni *et al.*, 1978).

tris(*N*-methylamino)methylsilane, a clear liquid at ambient temperature, can be synthesised by reaction of methylamine, CH_3NH_2, with methyltrichlorosilane, CH_3SiCl_3 at 313 K in petroleum ether (Penn *et al.*, 1982) according to the equation

$$CH_3SiCl_3 + 6CH_3NH_2 \rightarrow CH_3Si(NHCH_3)_3 + 3CH_3NH_3^+Cl^- \quad (7.4)$$

and polymerised at 793 K to a brittle transparent resin soluble in chloroform and methylene chloride. Gel-permeation chromatography showed an increase in weight-average molecular weight ($\bar{M}_w$) from 1537 to 4222 when the reaction time was extended from 1.5 to 4.5 h. Polymer was characterised by IR and NMR spectra and elemental analysis produced an atomic ratio $SiC_{2.78}N_{1.47}H_{9.15}$ corresponding to the general formula $[CH_3(CH_3NH)Si(CH_3N)]$ $[CH_3Si(CH_3N)_{1.5}]$ (Okamura, 1987). This polysilazane resin contained Si–N–Si linkages and may be compared with the structure of silazane monomers (figure 6.5). Fibres (12–100 μm diameter) drawn by hand from a polymer melt between 493 and 523 K and pyrolysed under N_2 at 1473 K resulted in a weight loss of 38%. The ceramic from this heat treatment was a homogeneous amorphous mixture of SiC/Si_3N_4 and had tensile strengths around 700 MPa, comparable with carbon fibre although its electrical resistivity was larger by a factor of 10^{12}. Transparent, amorphous oxynitride fibres, about 13 μm diameter, have been synthesised by heating melt-spun polycarbosilane ($\bar{M}_n = 1500$) in NH_3 between 1273 and 1673 K (Okamura *et al.*, 1984) and had an empirical formula $SiN_{1.5}O_{0.47}$ and tensile strength up to 1.8 GPa.

Complex chemical reactions which occur during synthesis of organosilicon ceramic precursors can be further illustrated by reference to hydridopolysilazane, HPZ (Legrow *et al.*, 1987). An exothermic reaction occurs on mixing trichlorosilane, $HSiCl_3$ and hexamethyldisilazane (figure 6.5) because of silicon–chlorine/silicon–nitrogen redistribution according to equations (7.5) and (7.6)

$$HSiCl_3 + ((CH_3)_3Si)_2NH \rightarrow (CH_3)_3SiNHSiHCl_2 + \\ + (CH_3)_3SiCl \quad (7.5)$$

$$(CH_3)_3SiNHSiHCl_2 + ((CH_3)_3Si)_2NH \rightarrow \\ ((CH_3)_3SiNH)_2SiHCl + (CH_3)_3SiCl \quad (7.6)$$

and these reactions are encouraged by distillation of $(CH_3)_3SiCl$ (boiling point 331 K) from the mixture. Further redistribution reactions can occur, for example

$$((CH_3)_3SiNH)_2SiHCl + ((CH_3)_3Si)_2NH \rightleftarrows$$
$$((CH_3)_3SiNH)_3SiH + (CH_3)_3SiCl \qquad (7.7)$$

but this reaction is slow and reversible. However, as the concentration of $((CH_3)_3SiNH)_2SiHCl$ increases an alternative thermodynamically favourable reaction occurs, namely

$$2((CH_3)_3SiNH)_2SiHCl \rightarrow ((CH_3)_3SiNHSiHCl)_2NH + $$
$$+ ((CH_3)_3Si)_2NH \qquad (7.8)$$

where disilazane can be separated from the tetrasilazane by distillation at 399 K. An additional reaction involves formation of NH_4Cl and a trisilyl-substituted nitrogen species

$$2((CH_3)_3SiNH)_2SiHCl \rightarrow$$
$$(CH_3)_3SiNHSiHClN((CH_3)_3Si)HSi(NHSi(CH_3)_3)_2 + HCl \qquad (7.9)$$

$$3HCl + ((CH_3)_3Si)_2NH \rightarrow 2(CH_3)_3SiCl + NH_4Cl \qquad (7.10)$$

which undergoes cyclisation and branching as polymerisation proceeds at 503 K. HPZ was obtained as a clear, colourless solid at ambient temperature, had a representative empirical formula $(SiH)_{39.7}((CH_3)_3 Si)_{24.2}(NH)_{37.3}N_{22.6}$ and $\bar{M}_w$ equal to 15 100. Its melt exhibited a decrease in viscosity from 51 Pa s at 463 K to 10 Pa s at 502 K and could be spun to fibres 10–15 μm in diameter: these viscosities lie within the range used for spinning Al_2O_3 fibres from aqueous sols (section 4.8). Curing was achieved by exposure to trichlorosilane vapour and, as for thermally cured polycarbosilane and cross-linked polysilastyrene, fibres were insoluble. A weight loss of 30% occurred at 1473 K in N_2 during decomposition to an amorphous ceramic that had tensile strengths up to 3 GPa and composition $(SiO_2)_{0.12}(Si_3N_4)_{1.00}(SiC)_{0.33}Si_{0.22}$. The general synthesis of polydisilylazanes from disilanes and hexamethyldisilazane has been described in more detail by Baney, Gaul & Hilty (1984).

Chlorosilanes allow synthesis of high-purity ceramics using liquid-phase reactions (chapter 6) and polymer pyrolysis because they can be readily purified by, for example, distillation. Organosilicon polymers can be pyrolysed at low temperatures compared with conventional routes to non-oxide ceramics such as direct nitridation of silicon because, as observed for alkoxide hydrolysis (chapter 5), interaction takes place on the molecular level before decomposition. A third advantage of these routes is that sintering aids, which can impair high-temperature mechanical properties, are not used during synthesis.

7.7 Composites, monoliths and coatings derived from polymer pyrolysis

The solubility and viscous nature of organosilicon polymers makes them suitable as binders for advanced ceramic components. For example, polysilane derived from $(CH_3)_2SiCl_2$ was reacted with $((CH_3)_3Si)_2NH$ in an autoclave at 723 K and the resulting solid polymer kneaded with SiC powder. The mixture could be pressed in a die to a green body (section 3.3) with the shape of a crucible that, after sintering at 1773 K, was subjected (Yajima, Hayashi & Omori, 1978c) to repeated cycles of infiltration and heat-treatment using the polymer solution; density and fracture strength of the ceramic were increased by this procedure. Infiltration can also be performed with molten polymer and binders are not restricted exclusively to organosilicon polymers. Hence the high-molecular-weight compound synthesised from $(CH_3)_2SiCl_2$, chromium acetate and hexacarbonyl molybdenum at 953 K contained 0.4 weight % Cr, 0.9 weight % Mo and produced higher density and strength than the polymer without these elements.

Polyborodiphenylsiloxane (section 7.4) has also been used as a binder (Yajima, Shishido & Hamano, 1977) whereby its benzene solution and SiC or Si_3N_4 powders were pressed at 196 MPa to bars that were densified at 1973 K without a dimensional change and with formation of a glassy B_4C phase. Techniques such as casting and injection molding allow fabrication of polymers into complex geometrical shapes difficult to obtain by hot-pressing.

Systematic studies on polymer pyrolysis have been made (Walker *et al.*, 1983; Coblenz *et al.*, 1984) and examples of materials and products are shown in figure 7.6. The carborane–siloxane is a liquid that can be thermally cross-linked at 573 K to a rubbery solid with elimination of benzene. In general, molecules that had cage, phenyl, unsaturated ring and unsaturated-chain structures gave higher yields on decomposition than precursors with saturated bonds and metal constituents contained in side groups (Walker *et al.*, 1983). Carborane–siloxane could be pyrolysed to small monolithic pieces while carbon substrates were infiltrated and dip-coated by this polymer and the polysilazane oil $(H_2SiNH)_i$ described in section 6.3. These non-oxide layers were, like alkoxide-derived coatings (figure 5.14), prone to crack but both the infiltration and surface-modification technique produced some increase in strength and oxidation resistance for substrate. This control of oxidation

111

Polymer pyrolysis

behaviour may be compared with enhanced oxidation resistance of porous SiC by infiltration with mixed $Si(OCH_3)_4/Ge(OC_2H_5)_4$ solutions (section 5.6).

Complex reactions between hexamethyldisilazane and chlorosilanes were outlined in equations (7.5)–(7.10) and analogous reactions have been used for synthesis of potential precursors for BN that contain

Precursor	Product	Yield (wt %)
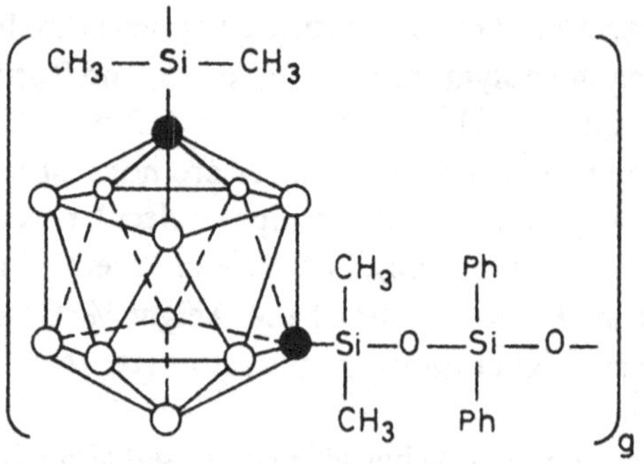 (i) Carborane–siloxane (o boron, ● carbon)	SiC/B₄C	60
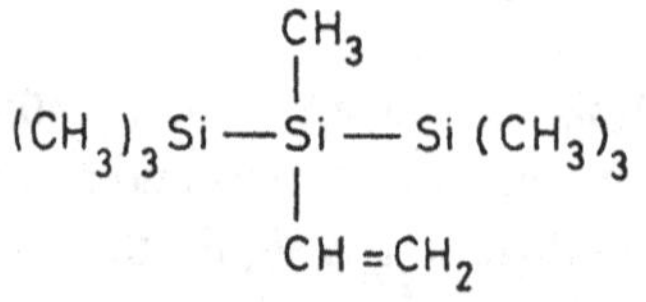 (ii) 1,1,1,2,3,3,3 – heptamethyl – 2 – vinyltrisilane	SiC	50
(iii) Ammonioborane H_3NBH_3	BN	65

Figure 7.6. Products from pyrolysis of ceramic precursors (Walker *et al.*, 1983).

112

Summary

B–N–B bonds (Narula, Paine & Schaeffer, 1986). Hence, B,B'-dimethyl B''-chloro N-trimethyl borazene reacts with $((CH_3)_3Si)_2NH$ to produce a cross-linked diborazene in high yield while heavy oils, soluble in methylene chloride and suitable for depositing films were obtained from combination of B-methyl B',B''-dichloro N-trimethyl borazene with heptamethyldisilazane, $((CH_3)_3Si)_2NCH_3$. Besides this liquid polymer, a granular solid could be synthesised from reaction of B-trichloroborazene with hexamethyldisilazane.

Polymeric ceramic precursors described in this chapter illustrate the role of chemistry at the organic–inorganic interface when applied to ceramic development and at the time of writing there is increasing interest in these synthetic pathways.

7.8   Summary

Pyrolysis of organosilicon polymers has attracted attention over the past 15 years because it enables synthesis of non-oxide ceramics that are difficult to prepare by other methods, particularly high-strength continuous silicon carbide fibre, which is now in industrial production. Besides fibre, monolithic pieces and coatings for enhanced oxidation resistance have been obtained while soluble polymers have potential application as binders for non-oxide components. The historical evolution of polymer pyrolysis shows how organosilicon materials, which initially had no applications in ceramics, became useful intermediates when their chemical properties and structures were elucidated.

8 Hydrothermal synthesis of ceramic powders

8.1 Introduction

Hydrothermal processing routes to ceramic powders involve heating reactants, often metal salts, oxide, hydroxide or metal powder as a solution or suspension in a liquid usually, but not necessarily, water at elevated temperature and pressure, up to *ca.* 573 K and *ca.* 100 MPa. Nucleation and particle growth that occur under these conditions can result in sub-micrometre oxide, non-oxide or metallic particles with controlled shape and size. Hydrothermal synthesis has not attracted the attention, as measured by the output of publications, that has been afforded to sol–gel processes described in chapters 4 and 5 even though it offers a direct route to monodispersed powders, an attractive property of advanced ceramic materials. Aspects of nucleation and growth processes that occur under hydrothermal conditions are described in this chapter together with examples of materials made by using the technique.

8.2 Forced hydrolysis of solutions at elevated temperatures and pressures

The hydrolysis of alkoxide solutions leading to monosized sub-micrometre oxide powders (section 5.7) requires that homogeneous nucleation occurs in one burst at a critical supersaturation concentration and particle growth takes place by diffusion of soluble species to the nuclei. The same principle applies to precipitation of particles from solutions at elevated temperatures and pressures (Matijević, 1981). Precipitation of metal hydroxides can be represented by the overall equation

$$M^{z+}(aq) + zOH^-(aq) \rightleftarrows M(OH)_z(s) \qquad (8.1)$$

although formation of intermediate soluble species prior to precipitation is given by the reaction

$$f[M(H_2O)_b]^{z+} + gOH^- \rightarrow$$
$$[M_f(H_2O)_{bf\text{-}g}(OH)_g]^{(fz\text{-}g)+} + gH_2O \qquad (8.2)$$

and it is these soluble species that are precursors to nuclei and affect particle growth. The reaction scheme shown in equation (8.2) involves introducing the base from an external source, but forced hydrolysis of electrolyte solutions at elevated temperatures leads to *in situ* formation of hydroxide ligands by deprotonation of bound H_2O molecules

$$f[M(H_2O)_b]^{z+} \rightarrow$$
$$[M_f(H_2O)_{bf\text{-}g}(OH)_g]^{(fz\text{-}g)+} + gH^+ \qquad (8.3)$$

This forced hydrolysis results in homogeneous nucleation, which can also take place following controlled release of precipitating anions or cations, the latter by decomposition of metal complexes (Matijević, 1981).

8.3 Hydrothermal reactions using salt solutions

Matijević and his group have carried out extensive studies on the hydrothermal synthesis of hydrous oxide sols. Monodispersed amorphous chromium hydroxide was obtained (Demchak & Matijević, 1969) from $CrK(SO_4)_2.12H_2O$, chromic sulphate and nitrate when the Cr(III) concentration was between 2×10^{-4} and 2×10^{-3} M, the solution had a pH < 5.4, contained SO_4^{2-} ions and was aged at temperatures of 333 K or higher; sols were produced in several days at 333 K and in about 18 h at 348 K. Particle diameters determined by light-scattering measurements on sols with a particle concentration of about 10^8 cm^{-3} ranged between 293 and 490 nm as the chrome alum molarity increased from 4×10^{-4} to 8×10^{-4} M. In the case of hydrous alumina (Brace & Matijević, 1973) monodispersed sols were prepared from aluminium sulphate, $AlK(SO_4)_2.12H_2O$ and aluminium nitrate containing Na_2SO_4 so that a typical reaction sequence involved ageing a solution 2×10^{-3} M in aluminium (pH 4.1) at 373 K for 84 h after which the pH had dropped to 3.1. Sulphate anions affected the particle size measured using light scattering and transmission electron microscopy, hence diameters increased from *ca.* 0.1 μm to 0.7 μm as the SO_4^{2-} concentration increased from 6.3×10^{-4} M to 4.0×10^{-3} M for a solution that was 10^{-3} M in Al. Sols with narrow size distribution were obtained in the titania system (Matijević, Budnik & Meites, 1977) by heating acidified solutions of $TiCl_4$ containing Na_2SO_4 for several weeks. A solution that was 0.16 M in Ti, 5.7 M in HCl and had a SO_4/Ti ratio between 0.47 and 3.8 underwent nucleation, which precipitated monosized crystalline rutile particles with diameters 1–4 μm when aged at 373 K for 41 days. Larger particle sizes resulted on prolonging the ageing time to between 15 and 45 days whereas diameters increased with the SO_4/Ti ratio for a constant heating period. In order to promote self-nucleation in these three oxide systems, solutions were filtered before use and heated in unstirred sealed tubes.

The work described above shows that the type of anion present in solution can affect formation of sol particles and their crystallinity. Sulphate ions were easily removed from amorphous hydrous alumina and chromium hydroxide by using a suitable washing procedure such as dialysis, but could also be incorporated into the crystalline lattice of alunite, $Fe_3(SO_4)_2(OH)_5.2H_2O$ made by heating acidified ferric sulphate solutions (Matijević, 1981). Hydrolysis products consisting of well-defined discrete ionic complexes such as those from ferric sulphate tend to yield particles with fixed stoichiometry and crystal habit but amorphous spherical particles are obtained when hydrolysis results in polymeric species. In these reactions, sulphate (or phosphate) can promote polymerisation through coordination with polynuclear species. Materials made by using forced hydrolysis are shown in figures 8.1 and 8.2.

Forced hydrolysis can be achieved by using controlled release of precipitating anions and an example is shown in figure 8.3 for a potential manufacturing route to Y_2O_3-stabilised zirconia powders in which the reactants are mixed solutions of yttrium chloride, zirconium oxychloride and $CO(NH_2)_2$. Decomposition of urea in an autoclave at temperatures in the range 433–493 K under pressures of 5–7 MPa releases ammonia, which precipitates hydroxides. Partially stabilised crystalline ZrO_2 obtained on raising the temperature was washed, dried, sintered and crushed to a superfine powder with particle size *ca.* 10 nm. An advantage of this hydrothermal pathway to sinterable zirconia powders is that low-quality reactants can be used because metallic impurities (e.g. Na) remain in solution. Also, variation of reaction conditions such as pH, temperature and pressure controls the particle size. The decomposition of urea is affected by pH. Initial products are NH_4^+ and NCO^-, which

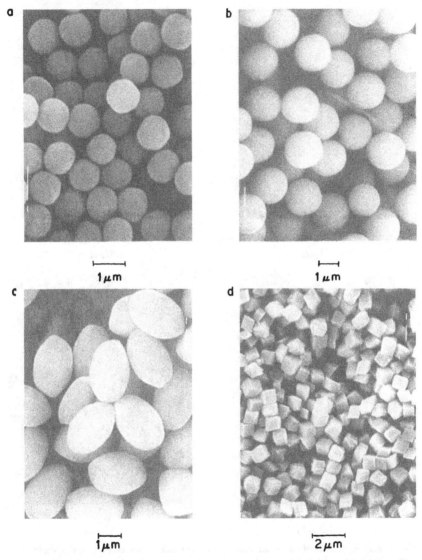

Figure 8.1. Examples of oxide, carbonate and sulphide particles made by forced hydrolysis (Matijević, 1984). (a) α-Fe$_2$O$_3$ by heating 0.04 M FeCl$_3$ and 0.005 M HCl at 398 K for 10 days. (b) CdS by ageing 0.0012 M Cd(NO$_3$)$_2$, 0.24 M HNO$_3$ and 0.005 M thioacetamide at 299 K for 14 h. (c) Iron(III) oxide by heating 0.019 M FeCl$_3$, 0.0012 M HCl and 5×10^{-5} M NaH$_2$PO$_2$ in a 50% ethanol/water solution at 393 K for 7 days. (d) Cadmium carbonate by mixing 10^{-3} M CdCl$_2$ with urea at room temperature.

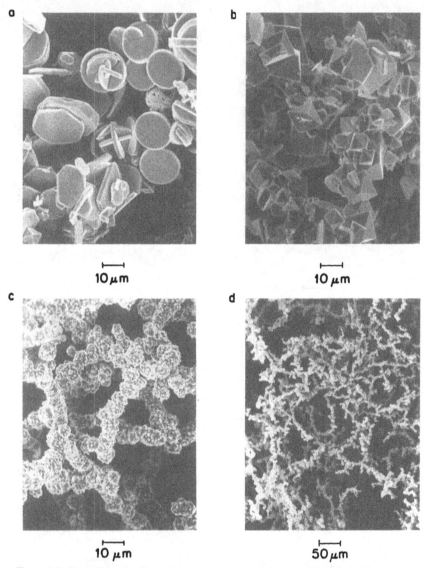

Figure 8.2. Examples of oxide and metal particles made by forced hydrolysis (Matijević, 1984). (a) α-Fe$_2$O$_3$ by heating a solution 0.04 M in Fe(ClO$_4$)$_3$, 0.2 M in triethanolamine, 1.2 M in NaOH and 0.5 M in H$_2$O$_2$ at 523 K for 1 h. (b) Fe$_3$O$_4$ by heating a solution 0.04 M in Fe(NO$_3$)$_3$, 0.2 M in triethanolamine, 1.2 M in NaOH and 0.85 M in N$_2$H$_4$ at 523 K for 1 h. (c) Cobalt by heating a solution 0.04 M in Co(II) acetate, 0.2 M in triethanolamine, 1.2 M in NaOH and 0.85 M in N$_2$H$_4$ at 523 K for 1 h. (d) Nickel by heating a solution 0.04 M in NiSO$_4$, 0.2 M in hydroxyethylethylenediaminetetraacetic acid, 1.2 M in NaOH and 0.85 M in N$_2$H$_4$ at 523 K for 1 h.

hydrolyses to CO_2 in acidic media, a reaction utilised for hydrothermal synthesis and also during internal gelation, described in section 4.6. However, NCO^- hydrolyses to CO_3^{2-} in alkaline conditions, allowing the preparation of metal carbonates, for example monodispersed

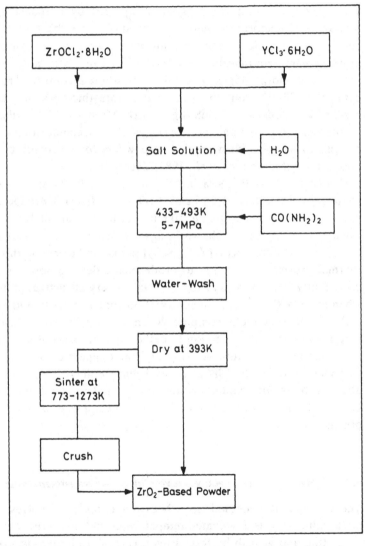

Figure 8.3. Schematic representation of a commercial process for producing zirconia powder by hydrothermal techniques (Anonymous, 1986).

$CdCO_3$ (figure 8.1d) made by mixing $CdCl_2$ with urea at ambient temperature (Matijević, 1984).

The controlled release of cations from metal complexes is the final method for preparing sols from solutions. An example is shown in figure 8.2a where Fe^{3+} ions complexed with triethanolamine, in the presence of NaOH and H_2O_2 are released into solution at 523 K leading to growth of disc-shaped α-Fe_2O_3; when H_2O_2 is replaced by a reducing agent, hydrazine, octahedral magnetite is formed (figure 8.2b). Metal chelates such as the complex made by using nickel acetate and hydroxyethyl ethylenediaminetetraacetic acid (HEDTA) can yield metal particles with various morphologies (figure 8.2c, d), whereas controlled release of cations at 299 K from the cadmium nitrate–thioacetamide complex results in monodispersed CdS (figure 8.1b; Matijević, 1987), illustrating a solution route to non-oxide ceramics. The decomposition of metal complexes has also been applied to powders for engineering ceramic applications (Stambaugh *et al.*, 1986). Hence nucleation from a solution (pH 7.4) that was 0.088 M in calcium nitrate, 0.176 M in zirconium oxynitrate, 0.5 M in NH_4OH and 0.176 M in HEDTA produced cubic zirconia with a particle size *ca.* 50 nm after heating at 466 K for 2 h. As for the reaction scheme shown in figure 8.3, impure reactants such as those derived from zircon ($ZrO_2.SiO_2$) sands can be used in this hydrothermal process. The ability to produce free-flowing powders of high crystallinity from cheap starting materials is very attractive, particularly when mixed oxides are homogeneous and contain little bound water.

A variation of metal-complex decomposition known as hydrolytic stripping involves precipitation of oxides by hydrolysis of non-aqueous metal carboxylate solutions using H_2O at temperatures up to 473 K (Monhemius & Steele, 1986). In this technique the oxide product is not affected by anion contamination and complex metal oxides (e.g. $CoFe_2O_4$) can be derived from mixed-metal complexes dissolved in the organic phase.

8.4   *Hydrothermal reactions involving phase transformations*

The experimental method described in section 8.3 involves heating metal-salt solutions at elevated temperatures and pressures to promote nucleation and growth by forced hydrolysis. A different class of reactions involves heating solid oxides or hydroxides that undergo phase transformations by dissolution and recrystallisation processes under hydrothermal conditions. Ferritic materials such as mixed Co,Ni oxides have industrial applications because of their magnetic properties, but also interest the nuclear industry because they represent a major corrosion product in water-cooled nuclear reactors. In order to synthesise chromium ferrites with Cr/Fe ratios up to 0.18 (Matijević *et al.*, 1986) mixed Cr, Fe hydroxides were aged as a suspension at 423 K for 16 h. After quenching reactants with cold H_2O, the solids were washed, dried and calcined between 613 and 673 K for 2 h in a hydrogen atmosphere, which resulted in black powders of about 1 μm diameter. Hydrothermal synthesis has been applied to the preparation of terbium-activated yttrium aluminium garnet (YAG) powders for use as phosphors in cathode-ray screens. The conventional method for phosphors (Hill, 1986) involves crushing and grinding powders followed by solid-state sintering. However, the luminescence effect is affected by particle size, contamination, size distribution and structural defects. Coarse particles reduce the resolution of the screen, whereas very fine particles of less than 1 μm cannot be excited as effectively as large ones. In the hydrothermal method (Takamori & David, 1986), mixed-hydroxide sol made by adding NH_4OH to chloride solutions of Al, Y and Tb was freeze-dried at 243 K to a fluffy powder with representative composition $Y_{0.39}Al_{0.56}Tb_{0.05}O_3$. Powder was aged in H_2O under 100 MPa in the temperature range 773–823 K for 15–20 h after pre-calcination between 673 and 973 K, which affected particle-size distribution during nucleation and growth. However, a heat treatment at 863 K resulted in monosized cubes of crystalline YAG with size *ca.* 5 μm after ageing at 773 K.

Hydrothermal synthesis involving phase transformations has also been used for preparation of tetragonal ZrO_2 for engineering ceramic applications (Tani, Yoshimura & Sōmiya, 1983). Amorphous ZrO_2 made by precipitation from $ZrCl_4$ solution using NH_4OH was calcined in air between 513 and 1093 K before ageing under 100 MPa at 473–873 K for 24 h in either H_2O or solutions of 30 weight % NaOH, 10 weight % KBr, 8 weight % KF and 7 weight %, 15 weight %, 30 weight % LiCl. Monoclinic and tetragonal ZrO_2 were obtained in H_2O, LiCl and KBr whereas only the monoclinic phase was produced in KF and NaOH; crystallite sizes derived from X-ray line broadening were between 15 and 40 nm. Particles, about 25 nm diameter, were isolated and non-aggregated in contrast with aggregates resulting from calcination of amorphous ZrO_2 in air, while their phase and crystallite size was a function of calcination temperature. The presence of amorphous oxide affected production of tetragonal ZrO_2, which was considered to occur by topotactic crystallisation on nuclei in the amorphous phase.

Reactions at elevated temperature and pressure have resulted in monodispersed particles of anatase 25–35 nm, rutile platelets 100–300 nm, monoclinic ZrO_2 10–32 nm, tetragonal ZrO_2 5 nm and ZrO_2. SiO_2 *ca.* 75 nm by using single-phase and diphasic gels derived from alkoxides and aqueous colloids (Komarneni *et al.*, 1986). Diphasic gels are considered (Hoffman, Roy & Komarneni, 1984) to consist, in for example the alumina–silica system, of discrete SiO_2-rich and Al_2O_3-rich regions or phases, whereas single-phase xerogel contains Al_2O_3 and SiO_2 mixed on an atomic scale in a non-crystalline structure. In the zircon system, single-phase gel was obtained on warming TEOS containing zirconyl chloride at 333 K, whereas diphasic gels were made from mixtures of aqueous colloidal SiO_2 and zirconia sols. Single-phase material gave zircon between 723 and 873 K under 100 MPa whereas diphasic gel resulted in mixtures of ZrO_2–SiO_2 and monoclinic ZrO_2 (baddeleyite) between 723 and 773 K but only zircon at higher temperatures. This was explained by independent reaction for ZrO_2 and SiO_2 phases in the diphasic materials (Komarneni *et al.*, 1986).

Ceramics for electronic applications have been synthesised by using the hydrothermal method. Calcium titanate was prepared (Kutty & Vivekanandan, 1987) by ageing at 423–473 K a hydrated titania slurry containing CaO derived from decomposition of calcium acetate at 973 K. An adsorbed layer of surface active agent, polyvinyl alcohol (molecular weight 8.5×10^4) modified the crystal morphology. In its absence, acicular crystals with irregular shapes several micrometres in length were produced but platelets 0.1–0.5 μm long with rectangular cross-section were obtained in its presence. Powders could be pressed and sintered to high-density pellets at 1673 K, compared with 1873 K for conventional milled $CaTiO_3$ materials, and reduced in 5% H_2/N_2 at 1373 K, which rendered them electrically conducting. Although dissolution of TiO_2 gel is unlikely because of low reaction temperatures, the mechanism for this hydrothermal process was considered to involve adsorption of Ca^{2+} onto the gel followed by cleavage of Ti–O–Ti bridging bonds accompanied by dehydroxylation leading to heterogeneous nucleation in the gel. Barium titanate substituted with Zr has also been prepared at low temperatures, 358–403 K, by the hydrothermal method (Vivekanandan, Philip & Kutty, 1986).

8.5 Hydrothermal reactions using metal reactants

The morphology of Co and Ni powders made by decomposition of metal complexes was shown in figure 8.2 c, d. However, oxides can be prepared by hydrothermal oxidation of metals, for example Cr_2O_3 (Sōmiya, 1984a) and ZrO_2 (Sōmiya, 1984b; Yoshimura & Sōmiya, 1982). In the case of hafnium (Toraya, Yoshimura & Sōmiya, 1983), metal chips (volume 4×10^{-6} cm³) were heated in water (H_2O/Hf mole ratio 2) under 50–150 MPa at temperatures in the range 573–973 K. Oxidation of Hf occurred in three stages. Between 573 and 673 K, HfO_2 was formed by surface oxidation

$$Hf + 2H_2O \rightarrow HfO_2 + 2H_2 \qquad (8.4)$$

but an oxide layer prevented oxidation in the interior of metal chips. However, hydrides such as $HfH_{1.5}$ were produced above 773 K by reaction of hydrogen with metal and at these temperatures H atoms diffuse rapidly through the solid. Hydrides were then oxidised by H_2O to monoclinic HfO_2 with particle size 20–40 nm, obtained from X-ray line broadening, and this reaction rather than surface oxidation produced the majority of oxide. Another hydrothermal reaction involving metals concerns reduction of metal hydroxides (Cu, Co, Ni, Ag, Pd and Os) in concentrated formic acid to the corresponding metals at temperatures between 453 and 513 K under pressure up to 25 MPa (Kutty *et al.*, 1982). Two general reactions were considered to occur under these hydrothermal conditions

$$M(OH)_2 + 2HCOOH \rightarrow M(HCOO)_2 + 2H_2O \qquad (8.5)$$

and

$$M(HCOO)_2 \rightarrow M + H_2O + CO + CO_2 \qquad (8.6)$$

although the exact reaction mechanism has not been fully elucidated.

8.6 Summary

Hydrothermal reactions involving homogeneous nucleation and phase transformations allow preparation of advanced ceramic materials such as stabilised zirconias in the form of isolated sub-micrometre oxide particles with controlled size, shape and chemical purity. The relatively low reaction temperatures, around 573 K, and ability to use impure reactants are attractive features for commercial exploitation of this technique.

9 Gas-phase reactions

9.1 Introduction

Gas-phase reactions are often associated with production in the laboratory and on an industrial scale of non-oxide ceramics such as silicon nitride. Reactants can be gaseous, volatile liquids or solids with reaction temperatures being as high as 15000 K. A common feature of powder preparations utilising gases, hydrothermal conditions and alkoxide hydrolysis is homogeneous nucleation, which was described qualitatively in sections 5.7 and 8.2. In this chapter, nucleation processes that give rise to liquid droplets from a supersaturated vapour are discussed, and the approach extended to ceramic particles with particular reference to control of particle size. A variety of experimental techniques involving flame hydrolysis, laser synthesis and plasma routes that have been used to prepare advanced ceramic powders are also described.

9.2 Gas-phase nucleation

The types of deposits that can arise from a gas-phase reaction are shown in figure 9.1. Whereas films, whiskers and bulk crystals are formed by heterogeneous nucleation on a solid substrate, particles are made by growth of nuclei following homogeneous nucleation.

Although the condition for mechanical equilibrium in a spherical surface is given by Laplace's formula (equation (5.12)) maintenance of both mechanical and physicochemical equilibrium for a system of liquid droplets in a vapour is described by the Kelvin equation (Defay et $al.$, 1966) namely

$$2\gamma v/r_d \mathbf{R}_G T = \ln (p'/p°) \qquad (9.1)$$

Gas-phase nucleation

where γ is the surface tension, v the molar volume of the liquid phase, and r_d the radius of a droplet with vapour pressure p' at temperature T. The saturation vapour pressure $p°$ corresponds to zero curvature ($p' = p°$ when $1/r_d = 0$), $p'/p°$ is the supersaturation and R_G the gas constant. Vapour pressure increases with a decrease in radius as shown in table 9.1 for water droplets.

The formation of spherical ceramic particles during a gas-phase reaction is governed by the same equations that apply for nucleation of liquid drops from a supersaturated vapour. The nucleation rate per unit volume, L, for liquid drops is given by the expression (Strickland-Constable, 1968)

$$L = \beta(p'/kT)^2 V_1(2\gamma/\pi m)^{1/2}\exp(-\Delta G'_n/kT) \qquad (9.2)$$

where β is an overall growth coefficient for droplets, V_1 the volume per molecule in the liquid, m the mass of a molecule and $\Delta G'_n$ the free energy of formation for a critical droplet or cluster containing n molecules, which may be written as

$$\Delta G'_n = 16\pi\gamma^3 V_1^2/3[kT\ln (p'/p°)]^2 \qquad (9.3)$$

Other variables in equations (9.2) and (9.3) have been previously defined. The effect of supersaturation on L for condensation of water

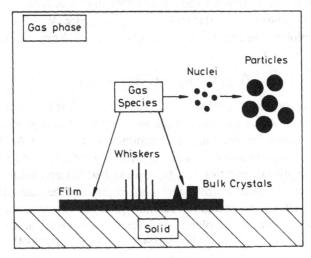

Figure 9.1. Schematic diagram illustrating the types of gas-phase deposits (Kato, Hojo & Okabe, 1981).

Table 9.1. *Effect of radius r_d
on the supersaturation $p'/p°$
for water droplets at 291 K
(Defay* et al., *1966)*

r_d/nm	$p'/p°$
∞	1
10^3	1.001
10^2	1.011
10	1.115
1	2.968

Table 9.2. *Effect of supersaturation on nucleation rate for water at 261 K
(Strickland-Constable, 1968)*

$p'/p°$	4	5	6
r_n/nm	0.93	0.81	0.72
n	114	73	53
L	10^{-10}	$10^{-0.7}$	$10^{4.2}$

droplets is shown in table 9.2 and particle growth proceeds after a cluster with critical radius r_n nucleates from the vapour phase. When a reaction between f moles of reactant F and g moles of G yields b moles of ceramic B and j moles of J

$$fF(g) + gG(g) \rightleftarrows bB(s) + jJ(g) \qquad (9.4)$$

the supersaturation is equal to $K(p_F^f p_G^g / p_J^j)$, where K is the equilibrium constant for the scheme shown in equation (9.4) and p_F, p_G, p_J are partial vapour pressures for reactants and products (Kato, Hoja & Watari, 1984). Hence high supersaturation is required to obtain powders from a gas-phase synthesis and this can be achieved with large values of K, whereas particle size is controlled by temperature and reactant composition, which affects the nucleation rate in equation (9.2). When a metal-containing reactant is completely converted to spherical powders the particle radius, a, is given by the expression (Kato *et al.*, 1984)

$$a = 0.5 \, (6C_0 M_w / N_p \rho)^{1/3} \qquad (9.5)$$

where N_p particles of ceramic, density ρ and molecular weight M_w are formed per unit volume from a metal source initially present at a concentration of C_0 moles per unit volume.

9.3   Flame-hydrolysed powders

The use of fumed or flame-hydrolysed oxide powders in sol–gel processing of colloids was outlined in section 4.8 and their purity described in table 5.3. These materials, for example SiO_2, are made by hydrolysis of $SiCl_4$ in a H_2/O_2 stationary flame (equation (5.11)) and particle formation is shown schematically in figure 9.2 (Ulrich, 1984). Molten spherical primary particles grow to larger droplets by coalescence. As particles begin to solidify they stick together, on collision, forming solid aggregates that also collide and form clusters known as agglomerates. The aggregated nature of Aerosil® silica, an oxide manufactured by using flame hydrolysis, is shown in figure 9.3. Synthesis of fumed oxide

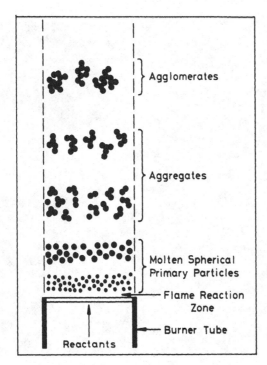

Figure 9.2. Schematic diagram illustrating the formation of primary particles, aggregates and agglomerates in a gas-phase reaction (Ulrich, 1984).

powders is related to the manufacture of optical fibres using the modified chemical vapour deposition (MCVD) method (Gambling, 1986). Mixed anhydrous chlorides such as $SiCl_4$, $GeCl_4$, BCl_3 and PCl_5 are oxidised at 1773–1873 K producing a homogeneous oxide coating on the inside of a cylindrical SiO_2 substrate. Because these chlorides are liquids they can be purified by distillation while refractive index gradients in the deposit are built-up on changing the composition of reactants. The silica tube is heated at about 2273 K when it collapses to a solid preform rod, which is drawn out to continuous fibre approximately 125 μm diameter.

9.4 Direct nitridation and carbothermic reduction

Ceramic synthesis using gases can be classified according to the heating method and physical state of reactants. This section is limited to reactions involving a solid phase and carried out in conventional resistance furnaces.

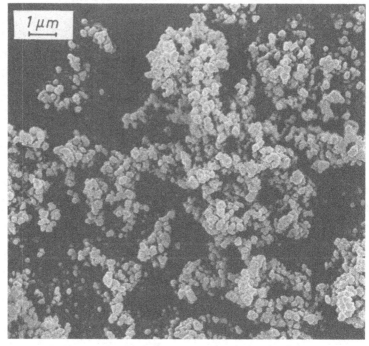

Figure 9.3. Scanning electron micrograph for a flame-hydrolysed Aerosil® silica (Courtesy of Degussa Company.)

Direct nitridation and carbothermic reduction

Silicon nitride is synthesised (Schwier, 1983) by direct nitridation, with controlled bed-depth, of chemically pure Si powder (less than 10 μm diameter) between 1473 and 1723 K in an atmosphere of NH_3, N_2/H_2 or N_2 according to the overall equation

$$3Si + 2N_2 \rightarrow Si_3N_4 \qquad (9.6)$$

This reaction is exothermic, and when N_2 is used the partial pressure of gas can be varied to prevent formation of $\beta\text{-}Si_3N_4$ (Lumby, 1976).

Silica is more readily available than silicon metal so that carbothermic reduction (Komeya *et al.*, 1984) of SiO_2 in a mixture containing fine SiO_2 and C powders followed by nitridation between 1473 and 1723 K in N_2 is an alternative route to $\alpha\text{-}Si_3N_4$

$$3SiO_2 + 6C + 2N_2 \rightarrow Si_3N_4 + 6CO \qquad (9.7)$$

The carbothermic pathway involves formation and reaction of gaseous silicon monoxide, SiO (Komeya & Inoue, 1975) as outlined below

$$3SiO_2(s) + 3C(s) \rightarrow 3SiO(s) + 3CO(g) \qquad (9.8)$$

$$3SiO(s) \rightleftarrows 3SiO(g) \qquad (9.9)$$

$$3SiO(g) + 3C(s) + 2N_2(g) \rightarrow Si_3N_4(s) + 3CO \qquad (9.10)$$

These two preparative methods produce $\alpha\text{-}Si_3N_4$ lumps that are comminuted and graded, operations that can introduce impurities into the ceramic. Additionally, excess carbon has to be used in the carbo-thermic method, which leaves a carbon residue. Although air oxidation removes the latter it also partially oxidises Si_3N_4 to SiO_2, whose presence can have deleterious effects on densification and integrity of sintered structural components. Silica and carbon powders are usually dry-mixed but more intimate mixing of components is achieved by using aqueous silica sols and tetraethyl silicate as described in sections 4.8 and 5.5, respectively. Properties of Si_3N_4 powders manufactured by direct nitri-dation and carbothermic reduction are listed in table 9.3; a scanning electron micrograph for Si_3N_4 (LC12 grade) made by nitridation of Si is shown in figure 9.4. This powder has a surface area of 20 $m^2\,g^{-1}$ and con-tains 94 weight % $\alpha\text{-}Si_3N_4$, 3 weight % $\beta\text{-}Si_3N_4$, 0.1 weight % free Si, 0.03 weight % Fe, 0.1 weight % Al, 0.02 weight % Ca and 0.005 weight % Na.

Carbothermic reduction is not restricted to silicon nitride. When heated in a continuous furnace, powder blends of TiO_2, C and boron

Table 9.3. *Properties of commercially available silicon nitride powders (Segal, 1986)*

Preparative method	Carbothermic reduction of SiO_2 in N_2	Nitridation of Si in N_2	Vapour-phase reaction of $SiCl_4/NH_3$
Manufacturer	Toshiba	H.C. Stark	Toya-Soda
Grade	—	HI	TSK TS-7
Metallic impurities (weight %)	0.1	0.1[a]	0.009[c]
Non-metallic impurities (weight %)	4.1	1.7[b]	1.2[d]
α-Si_3N_4 (weight %)	88	92	90[e]
β-Si_3N_4 (weight %)	5	4	10
SiO_2 (weight %)	5.6	2.4	—
Surface area/($m^2\,g^{-1}$)	5	9	12
Grain size/μm	0.4–1.5	0.1–3	0.2–3
Tap density/($kg\,m^{-3}$)	430	640	770

[a] Fe (0.04); Al (0.10); Ca (0.03) (expressed as weight %).
[b] O (1.3); C (0.4).
[c] Fe (0.005); Al (<0.001); Ca (<0.001); Na (<0.001); K (<0.001).
[d] O (1.0); C (0.1); Cl (0.1).
[e] After 1673 K calcination.

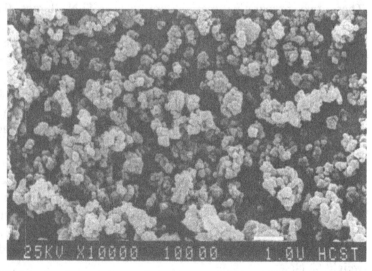

Figure 9.4. Scanning electron micrograph for a commercially available silicon nitride powder (LC12 grade) made by nitridation of silicon. (Courtesy of Hermann C. Starck, Berlin.)

carbide (98.5 weight % purity) yield titanium diboride 5–10 μm diameter after milling with chemical purity greater than 98 weight % (Sheppard, 1987). Limitations on drawing oxide fibres from a melt were discussed in section 4.8. However, melt-spun B_2O_3 fibres (6 μm diameter) can be nitrided in ammonia according to the overall equation

$$B_2O_3 + 2NH_3 \rightarrow 2BN + 3H_2O \qquad (9.11)$$

although the reaction is more complex than indicated and occurs in three stages (Economy, Smith & Lin, 1973)

$$bB_2O_3 \text{ (fibre)} \xrightarrow[NH_3]{>473 \text{ K}} (B_2O_3)_b.NH_3 \qquad (9.12)$$

$$(B_2O_3)_b.NH_3 \xrightarrow[NH_3]{>623 \text{ K}} (BN)_f(B_2O_3)_g.(NH_3)_j + H_2O \qquad (9.13)$$

$$(BN)_f(B_2O_3)_g.(NH_3)_j \xrightarrow{>1773 \text{ K}} BN \text{ (fibre)} + \\ + B_2O_3 + H_2O + NH_3 \qquad (9.14)$$

where b, f, g and j have been previously defined. Reactions (9.12) and (9.13) were rate-determining, in which $NH_3(g)$ diffused into oxide fibre producing the addition complex shown in equation (9.12) while at temperatures greater than 623 K substitution occurred releasing H_2O, a process probably involving mixtures of cyclic BN and boroxine rings, which coalesced to a disordered three-dimensional layer-type BN structure; nitrogen content of the fibre increased during this step. Equation (9.14) represented a fast reaction completed in several seconds that produced white filamentary boron nitride, which could be fabricated into various shapes including felts and papers.

9.5 Non-plasma gas-phase reactions

$SiCl_4$ and SiH_4 react with NH_3 according to the overall equations

$$3SiCl_4(g) + 4NH_3(g) \rightarrow Si_3N_4(s) + 12HCl(g) \qquad (9.15)$$

$$3SiH_4(g) + 4NH_3(g) \rightarrow Si_3N_4(s) + 12H_2(g) \qquad (9.16)$$

The use of $SiCl_4$ leads to corrosive by-products, whereas SiH_4 is chemically more dangerous than the tetrachloride because of its spontaneous flammability in air. Prochaska & Greskovich (1978) reacted SiH_4 with NH_3 at atmospheric pressure between 773 and 1173 K and obtained Si_3N_4 powders with particle diameters in the range 30–200 nm, nitrogen surface areas up to 26 m^2 g^{-1}, cation impurities less than 100 p.p.m. and oxygen content less than 2 weight %, although the NH_3/SiH_4 ratio affected crystallisation temperature (up to 1753 K). Properties of a commercial powder made by reaction between $SiCl_4$ and NH_3 are shown in table 9.3. Chlorine retention in nitride powders can have adverse effects on the mechanical properties of sintered components and the synthesis of Cl-free Si_3N_4 involving the Si–S–N system is described in section 9.7.

An alternative to furnace heating has been pioneered by Haggerty & co-workers at the Massachusetts Institute of Technology who used a 150 W CO_2 laser as the heat source. In their work (Cannon *et al.*, 1982a, b), a cross-flow cell was used whereby reactant gases diluted in a carrier stream (argon) emerged from a nozzle and orthogonally intersected the laser beam. Silane was used in preference to $SiCl_4$ because of a strong absorbance band near the wavelength, 10.6 μm, of the laser radiation. Excitation of all gaseous components by the laser is unnecessary because absorbance time for photons is greater than the time required for distribution of energy by molecular collisions. Si_3N_4 was prepared by using the gases shown in equation (9.16) whereas Si and SiC powders were made according to the reactions

$$SiH_4(g) \rightarrow Si(s) + 2H_2(g) \qquad (9.17)$$

$$2SiH_4(g) + C_2H_4(g) \rightarrow 2SiC(s) + 6H_2 \qquad (9.18)$$

Fast heating and cooling rates of 10^6 K s^{-1}, 10^5 K s^{-1} respectively, together with reaction times around 10^{-3} s, are characteristic properties for these laser-driven processes (Haggerty *et al.*, 1986), which yield powders at the rate of about 10 g h^{-1}, while gas pressure, reactant-gas ratios, mass-flow rates and laser intensity control particle size, crystallinity and stoichiometry of the powders (Kizaki, Kandori & Fujitani, 1985; Symons, Nilsen & Danforth, 1986). Representative processing conditions are shown in table 9.4, and table 9.5 contains powder properties. Stoichiometric Si_3N_4 contains 60 weight % Si and 40 weight % N, which indicates that material in table 9.5 is deficient in nitrogen although stoichiometric silicon nitride was obtained on increasing the laser intensity (Cannon *et al.*, 1982b); nitride powders were theoretically dense and crystallised in N_2 between 1573 and 1673 K. The oxygen content of Si_3N_4 powders derived from laser-heated gases (table

Table 9.4. *Representative processing conditions for laser-driven gas-phase reactions (Cannon et al., 1982a)*

Powder	Si	Si_3N_4
Laser intensity/(W cm^{-2})	760	760
Pressure/Pa	2×10^4	2×10^4
Gas velocity at nozzle/(m s^{-1})	1.04	11.4
Gas velocity at laser beam/(m s^{-1})	0.36	9.49
Reaction-zone temperature/K	1303	1140
Gas-flow rate/(cm^3 min^{-1})		
(i) SiH_4	11	11
(ii) NH_3	—	110
(iii) Ar	1000	1000

Table 9.5. *Physical properties of powders derived from laser-driven gas phase reactions (Cannon et al., 1982a, b)*

Powder	Si	Si_3N_4	SiC
Particle size from gas adsorption/nm	45.8	17.6	24.7
Mean particle size from electron microscopy/nm	47.0	16.8	23.0
Size range from microscopy/nm	10–50	10–25	—
Surface area/(m^2 g^{-1})	57	117	97.5
Crystallinity	Crystalline	Amorphous	Poorly crystalline
Si (weight %)	99	72	80
N (weight %)	0.02	26.0	—
O (weight %)	0.06	0.05	1.4
C (weight %)	—	—	14.1
Free Si (weight %)	100	35	—

9.5) is less than in Si_3N_4 made by other methods (table 9.3); lower O_2 and hence SiO_2 levels are desirable for sintering and strength of components.

Heterogeneous nucleation is absent because the laser does not heat the reactor walls, while processing conditions encourage development of uniform nucleation and growth. Properties required for an ideal sinterable powder were outlined in section 3.9, namely small particle size, a narrow range of sizes to avoid grain growth, an absence of aggregates, sphericity for improved powder packing and control of chemical purity, conditions that are met by laser-driven reactions. Small particles about 20 nm diameter nucleate directly from the gas phase whereas weakly bound aggregates can arise because of van der Waals forces. Larger Si particles about 0.2 μm diameter that are shown in figure 9.5 were synthesised from SiH_4 by using a total gas pressure of 0.15 MPa (Haggerty *et al.*, 1986) in which the temperature exceeded the melting point of Si (1683 K) allowing growth by coalescence of molten particles.

Synthesis of silicon nitride from Si powder was described in section 9.4. Silicon metal is manufactured by carbothermic reduction of SiO_2 but has to be comminuted and chemically purified before nitridation.

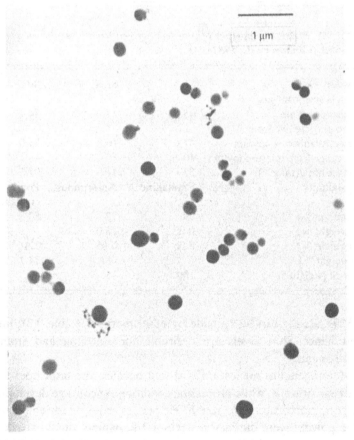

1 μm

Figure 9.5. Transmission electron micrograph for silicon powder made by a laser-driven gas-phase reaction (Haggerty *et al.*, 1986).

Industrial SiC powders are made in the Acheson process by carbo-thermic reduction of SiO_2 using resistance furnaces at temperatures in excess of 2073 K (Kennedy & North, 1983) as the reaction is endothermic involving generation of gaseous SiO. The commercial product has a large grain size and is contaminated with oxygen. Hence laser-driven syntheses are potentially attractive routes to non-oxide ceramic powder if scale-up does not incur significant cost penalties.

Reaction-bonded silicon nitride (RBSN) has been prepared from laser-synthesised Si powder. In early studies (Danforth & Haggerty, 1983), dry pellets made by cold pressing at 6.9 MPa followed by isostatic pressing at 310 MPa to greater than 70% of theoretical density were sintered at 1623 K under 0.1 MPa. Nitridation was performed in a 10% H_2/N_2 atmosphere for 20 min at temperatures up to 1623 K and fracture strengths for nitrided compacts, which had densities between 57 and 88% of theoretical value, were in the range 156–352 MPa. Later work by Haggerty and his co-workers, (1986) showed that higher fracture strengths could be obtained when powder was initially dispersed to a non-aqueous Si sol in CH_3OH that broke down aggregates after which compacts were made by collecting sol particles on a filter. Powder cakes were further densified under 69 MPa and then nitrided for 1 h at 1693 K, which resulted in an α/β phase ratio around 90/10, a density 76% of the theoretical value and maximum strength of 450 MPa that was higher than for other RBSN samples. This example illustrates how colloid chemistry can be applied to a filtration technique and the control of powder packing that may be obtained by using a dispersion.

Chemical syntheses described in this section have involved SiH_4 and $SiCl_4$ as sources of Si but an alternative reactant, hexamethyldisilazane (figure 6.5), has been used (Rice, 1986). Vapour from this liquid silazane was diluted in C_2H_4 or NH_3 and injected into the laser beam using a cross-flow geometry with total gas pressure in the range 5.6×10^4–6.7×10^4 Pa and laser flux between 145 and 480 W cm^{-2}. The resulting agglomerated powders (primary particle size 100 nm), which had a representative composition $SiN_{0.45}C_{1.2}H_{0.5}$, did not sinter well and yielded SiC on calcination. Si–N bonds are stronger than Si–C bonds, resulting in preferential cleavage of the latter. Unlike silane decomposition, which produced a supersaturated vapour of Si atoms, the silazane dissociated to small organic molecules, radicals and large organosilicon molecules, for example $(CH_3)_3SiNHSi(CH_3)_2NHSi(CH_3)_3$, which were precursors for particle growth.

9.6 Plasma reactions

Considerable efforts have been made since the early 1950s on the development of direct current (d.c.), alternating current (a.c.) and radio-frequency (r.f.) thermal plasmas, which have found applications in a wide range of industrial processes. Thermal-arc plasmas were used initially for welding and cutting whereas r.f. plasmas had been developed for novel crystal-growth techniques (Fauchais *et al.*, 1983). They have been used for decomposition of industrial wastes, spheroidisation of materials, coal desulphurisation, extractive metallurgy and mineral exploitation (Kong & Pfender, 1987). The deposition of coatings by plasma spraying was outlined in section 4.8, whereas silicon nitride films have been produced at low temperatures, about 573 K, by plasma-activated vapour deposition onto semiconductor devices for applications as scratch-resistant layers and moisture barriers on integrated circuits (Goodwin, 1982). This section, however, is restricted to the synthesis of advanced ceramic powders in plasmas. These reactions offer potential advantages for powder preparation, fast reaction times due to high temperatures and rapid cooling rates leading to a high degree of supersaturation and homogeneous nucleation.

Canteloup & Mocellin (1975) developed an r.f. argon plasma reactor operating at 8.2 MHz and a 10 kW power level in which vapours derived from evaporation of Si (particle size less than 35 μm) and Al powders injected into the tail of the plasma flame reacted with N_2 or NH_3 at temperatures as high as 6000 K; metal reactants avoided production of corrosive gases while residence time in the flame was about 1.4×10^2 s. A low efficiency (15%) for conversion to a mixture of α- and β-Si_3N_4 containing unreacted Si occurred in N_2, but much higher yields were obtained in NH_3; similar efficiencies resulted with Al. Particle diameters for AlN were 7–8 nm corresponding to surface areas of 260 m^2 g^{-1} although larger sizes, 11–14 nm, were produced for Si_3N_4 powders that had surface areas around 160 m^2 g^{-1}.

Vogt and his co-workers have carried out systematic studies on the synthesis, in thermal plasmas, of SiC, Si_3N_4, Al_2O_3, Ni and B_4C powders (Vogt *et al.*, 1984, 1985). Their reactor, shown schematically in figure 9.6, consists of a copper induction tube (7.5 cm diameter) housed in a quartz mantle that was connected to a water-cooled stainless-steel quenching chamber (19.5 cm diameter); typical operating parameters were 360 kHz, a power level between 25 and 40 kW and gas pressure of

8×10^4 Pa. Al_2O_3 was prepared by oxidation of aluminium trimethyl whereas Ni powders were synthesised from $Ni(CO)_5$. Unlike a.c. systems, the absence of electrodes in r.f. heated gases reduces contamination in ceramic powders. Aggregated β-SiC, with surface area ranging from 60 to 130 m^2 g^{-1} and primary particle size 10–60 nm, was synthesised from SiH_4/CH_4 mixtures. Amorphous Si_3N_4 was obtained for NH_3/SiH_4 ratios greater than 5.5 whereas a mixture of α- and β-phases containing 50 mole % unreacted Si resulted when this ratio was 1.6–3.8. Zhu & Yan (1985) also made Si_3N_4 in a plasma but used $SiCl_4$ rather than SiH_4 as the silicon source. Plasma-derived nickel powders

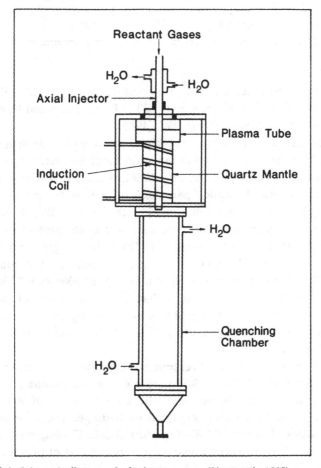

Figure 9.6. Schematic diagram of r.f. plasma reactor (Vogt *et al.*, 1985).

Gas-phase reactions

Table 9.6. *Comparison of chemical purity between plasma-derived and commercial silicon carbide powders (Carbone & Rossing, 1986)*

Element	Plasma powder (p.p.m.)	Commercial powder (p.p.m.)
Al	<30	300–500
B	—	<5–200
Ca	<20	<20–200
Cu	<20	<30–100
Fe	<10	100–400
Ti	<100	40–200
W	<300	<300–1000
O	10^3–3×10^3	8×10^3–19×10^3
Free C	10^3–15×10^3	3×10^3–18×10^3
Cl, F	300–500 (Cl)	<1200 (F)

20 nm–3 μm diameter sintered to theoretical density at 1073 K in H_2 compared with temperatures about 1223 K for conventional Ni powder with a 2–5 μm size range (Vogt *et al.*, 1985).

Higher chemical purity can be obtained by using plasmas than in conventional synthesis, as shown in table 9.6 for β-SiC made by reacting $SiCl_4$ with CH_4 (Carbone & Rossing, 1986). A similar powder consisting of both equiaxed and spherical particles 0.6 μm diameter and containing 0.17 weight % boron as a sintering aid introduced as BCl_3 to reactant gases was initially uniaxially pressed, then isostatically pressed under 138 MPa and finally densified in vacuum at 2223–2448 K to greater than 97% theoretical density (3210 kg m^{-3}). Larger amounts of boron (0.23 weight %) were required to produce a density of 3060 kg m^{-3} for conventional β-SiC powder. The enhanced sinterability of plasma-synthesised material was assigned to greater homogeneity for B and Si constituents and lower O_2 levels (0.17 weight %) than in conventional β-silicon carbide.

Plasma routes to advanced ceramic powders are not restricted to r.f. sources. Kong & Pfender (1987) used a d.c. argon-plasma jet reactor operating at atmospheric pressure with an arc current of 900 A and voltage 25 V in which SiO generated by hydrogen reduction of SiO_2 reacted with CH_4 above 10^4 K to produce β-SiC. Cooling rates greater than 10^6 K s^{-1} encouraged homogeneous nucleation of powder in the size range 2–40 nm with a 97.3% conversion efficiency.

138

Plasma synthesis is also applicable to multicomponent systems. Hence 24 weight % Cr_2O_3–Al_2O_3 has been made by oxidation of halides, namely $AlBr_3$ and CrO_2Cl_2 vapours in an oxygen–argon plasma (McPherson, 1973) and SnO_2–Al_2O_3 containing up to 24.5 weight % SnO_2 from $AlBr_3$/$SnCl_4$ mixtures (Gani & McPherson, 1987). Liquid droplets that condense in the plasma grow by coalescence, and figure 9.7 shows a δ-Al_2O_3 powder prepared in this way. The work was extended to doped ZrO_2 powders using $ZrCl_4$, $TiCl_4$, $AlCl_3$, $SiCl_4$ and CrO_2Cl_2 in a 15 kW, 5 MHz plasma torch (McPherson & Aik, 1987). Compositions containing up to 70 weight % ZrO_2 in the Al_2O_3–ZrO_2 system, relevant to the preparation of toughened alumina, consisted of δ-Al_2O_3 and tetragonal zirconia. Plasma reactors and transport properties of plasmas have been described in the excellent review by Fauchais *et al.* (1983).

9.7   The silicon–sulphur–nitrogen system

It was mentioned earlier in this chapter that impurities, for example oxygen and Cl in α-Si_3N_4, can affect mechanical properties, particularly

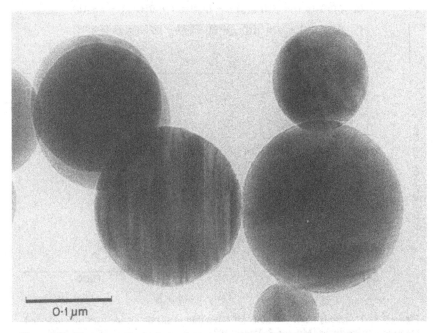

0·1 μm

Figure 9.7. Plasma-synthesised alumina powder (Gani & McPherson, 1974).

139

those at high temperature, of sintered components. Studies by Morgan (1980, 1983) together with Pugar (1985a, b) illustrate a novel chemical route to pure silicon nitride. Calculated free-energy changes for conversion of $SiCl_4$ and silicon disulphide, SiS_2, to Si_3N_4 are shown in figure 9.8. While $SiCl_4$ reacts with a 6 volume % H_2/N_2 mixture (producer gas) at 1700 K without formation of NH_4Cl, which can lead to chlorine retention, it also reacts with alumina furnace tubes as indicated in the equation

$$3SiCl_4 + 2Al_2O_3 \rightarrow 3SiO_2 + 4AlCl_3 \qquad (9.19)$$

Similar reactions occur between Al_2O_3 and SiI_4 (Morgan & Pugar, 1985a). Conversion of SiS_2 to silicon nitride in dry ammonia (figure 9.8) is more favourable thermodynamically than the reaction shown in equation (9.19). In addition, Si–S is a weaker bond than Si–Cl (table 9.7), while silicon monosulphide, SiS, has a higher vapour pressure and stability than SiO in the gaseous state (Morgan & Pugar, 1985a). Silicon

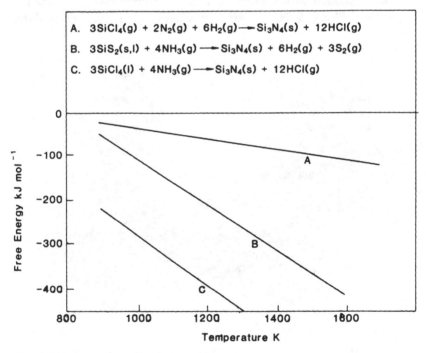

Figure 9.8. Thermodynamic relations for formation of silicon nitride from sulphide and chloride intermediates (Morgan & Pugar, 1985a).

Table 9.7. *Bond strength for the silicon halide, nitrogen, oxygen and sulphur systems (Morgan & Pugar, 1985a)*

Bond	Bond strength/(kJ mol^{-1})
Si–Cl	380
Si–I	213
Si–N	310–330
Si–O	435
Si–S	330

disulphide with a needle or whisker morphology was obtained after passing 10 volume % H$_2$S/Ar over powdered Si at 1173 K

$$\text{Si} \xrightarrow{\text{H}_2\text{S}} \text{SiS}_2 + \text{H}_2 \qquad (9.20)$$

and grew by a vapour–liquid–solid mechanism in which eutectic droplets absorbed SiS and S vapour. The polymeric Si–S–N–H intermediate obtained on reaction of NH$_3$ with SiS$_2$ between 298 and 1173 K

$$\text{SiS}_2 \xrightarrow{\text{NH}_3} [\text{Si–S–N–H}]\,\text{polymer} \qquad (9.21)$$

could be converted in NH$_3$ at 1173–1723 K to amorphous silicon nitride, which had a 'spaghetti-shaped' morphology (figure 9.9) and was free of sulphur (Morgan, 1983).

9.8   Electron-beam evaporation

The synthesis of ceramic powders from solids in conventional resistance furnaces was described in section 9.4. Electron-beam evaporation of solids was developed by Ramsay & Avery (1977) as the heating method for preparing ultrafine B$_4$C and SiC with particle size less than 100 nm. Hence when boron carbide granules were subjected to an electron beam in a methane atmosphere at 133 Pa, their surface temperature increased to about 2773 K and evaporation of solid occurred at the rate of 10 g kWh^{-1}. Vapour condensed to a fluffy dark-brown powder with nitrogen surface area 381 m^2 g^{-1} and corresponding particle size of 6.3 nm. These

Gas-phase reactions

powders could be hot-pressed to compacts at lower temperatures than for conventional large-grain material.

9.9 Summary

A variety of heating methods, conventional resistance furnaces, lasers, plasmas and electron beams have been used to carry out gas-phase reactions involving solid, liquid and gaseous reactants. The techniques often involve homogeneous nucleation and have been particularly applied to non-oxide ceramic powders such as silicon nitride on the laboratory and industrial scale. Powders can be obtained in high purity with a sub-micrometre particle size and narrow size distribution, desirable properties for sintering advanced ceramic materials.

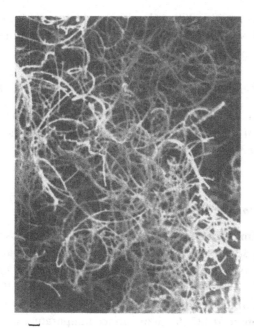

1 μm

Figure 9.9. Amorphous 'spaghetti-shaped' silicon nitride derived from silicon disulphide (Morgan & Pugar, 1985a).

142

10 Miscellaneous synthetic routes to ceramic materials

10.1   Introduction

The previous six chapters have been concerned with non-conventional preparative routes to ceramic materials, that is, techniques which avoid powder mixing and milling or coprecipitation of hydroxides and oxalates. Synthetic routes not associated with sol–gel processing of both colloids and metal–organic compounds, non-aqueous liquid-phase reactions, polymer pyrolysis, hydrothermal synthesis and gas-phase reactions are the subject of this chapter. These methods are the citrate gel process, alkoxide pyrolysis, freeze-drying and rapid expansion of supercritical solutions. Different drying processes, which have been mentioned in earlier sections, are discussed here with particular reference to their role in modifying the sintering behaviour of ceramic powders.

10.2   The citrate gel process

The citrate gel process, which was developed by Marcilly & co-workers (1970), can be illustrated by the synthesis of Gd_2O_3-stabilised zirconia (van der Graaf & Burggraaf, 1984). In order to form an organometallic complex, citric acid was added to mixed zirconyl and Gd(III) nitrate solutions (acid/metal mole ratio = 2:1) whose pH had been increased with NH_4OH to between 6 and 7.5. Other organic acids containing at least one hydroxyl and one carboxylic group such as tartaric, lactic and glycollic acid can be used. Rapid partial dehydration produced highly viscous mixtures that could be dried to amorphous gels at 373 K. The viscous ceramic precursor swelled on further heating because of decomposition of NH_4NO_3 after which the organometallic complex decomposed exothermically. Stabilised zirconia from this reaction exhibited a tissue-like morphology when calcined at 923 K and had a surface area of 58 m^2

g^{-1}. Cations are contained in the dry gel as a homogeneous mixture, which results in oxides with uniform composition on the molecular scale. This technique was used initially for barium hexaferrite, $BaFe_{12}O_{19}$, lanthanum chromite, $LaCrO_3$ and $MgAl_2O_4$ (Marcilly *et al.*, 1970).

The citrate gel process has recently been extended (Mahloojchi *et al.*, 1987; Tang *et al.*, 1987) to high-T_c oxide superconductors with composition $YBa_2Cu_3O_{7-\delta}$ and $La_{1.7}Sr_{0.15}Ba_{0.15}CuO_4$ in order to obtain greater homogeneity in and reproducibility for the oxide than is obtainable by mixing powders. It should be mentioned that although the amorphous solids are loosely referred to as gels they are not identical to gels derived from aqueous colloids and alkoxide solutions.

A method related to the citrate gel process uses an organic acid complexing agent for synthesis of sub-micrometre $BaFe_{12}O_{19}$ powder with high coercivity (Lucchini *et al.*, 1984). Barium nitrate solution was mixed with the ammonium salt of polygalacturonic acid, which resulted in cation binding on the molecular chain of this acid. Ferric nitrate solution was then added and the mixture decomposed, after freeze-drying, to hexaferrite at 973 K, a temperature considerably lower than those in excess of 1273 K that are required for obtaining ferrite at a significant rate in the industrial process by using Fe_2O_3 and $BaCO_3$ as starting materials. The rate of formation for $BaFe_{12}O_{19}$ was controlled by oxidation of Fe^{2+} ions produced on combustion of the organic acid.

10.3 Pyrolysis of metal alkoxides

The preparation of sub-micrometre oxide powders using alkoxide hydrolysis was described in chapter 5 although similar materials have been synthesised by pyrolysis of alkoxides, for example $Zr(O'C_4H_9)_4$ in N_2 at 673 K (Mazdiyasni, 1982). While high-purity (more than 99.95 weight %) oxides result from the inherent purity of starting reagents (table 5.2) these powders were characterised by a very small primary particle size, about 1 nm. However, powders were agglomerated with representative mean size of 49 nm after heating at 1223 K and it emerged that a requirement for obtaining small particles was rapid and quantitative hydrolysis. Products from $Zr(O'C_4H_9)_4$ were olefin, alcohol and ZrO_2 so that although decomposition could be represented by the overall equation

$$Zr(OR)_4 \rightarrow ZrO_2 + 2ROH + \text{olefin} \qquad (10.1)$$

the suggested mechanism was

$$Zr(OC_4H_9)_4 \rightarrow Zr(OC_4H_9)_3OH + (CH_3)_2C(CH_2) \qquad (10.2)$$

$$Zr(OC_4H_9)_3OH \rightarrow ZrO_2 + 2C_4H_9OH + (CH_3)_2C(CH_2) \qquad (10.3)$$

which may be compared with the reactions described by equations (5.18)–(5.20) for hydrolysis of $Ti(OC_2H_5)_4$.

10.4   *Rapid expansion of supercritical solutions*

Preparation of aerogels and monoliths using supercritical drying conditions was described in section 5.5, although in a recently developed technique (Matson, Petersen & Smith, 1986a, b) ceramic powders and fibres have been synthesised by rapid expansion of supercritical solutions (RESS). In the RESS method, solute nucleation and condensation take place within an expanding supercritical jet following the abrupt loss of solvating power when fluid is transferred through a nozzle. For example, solutions containing soluble SiO_2 species with concentration up to 3000 p.p.m. at 743 K and under 60 MPa were expanded through a stainless-steel nozzle 60 μm in diameter with flow rates around 0.7 cm^3 s^{-1} (Matson *et al.*, 1986a). Higher concentrations produced by raising the autoclave temperature increased the particle diameter whose maximum value was 2 μm corresponding to 673 K, although particle size was also affected by the temperature at the nozzle during expansion. Low electrolyte concentrations (10^{-4}–10^{-3} M NaCl) reduced diameters but enhanced agglomeration between particles, possibly because of the effect of ionic strength on colloid stability (section 4.3).

The RESS process has also been applied to powders and fibres of polycarbosilane (Matson *et al.*, 1986b). Expansion of pentane solutions containing between 35 and 3500 p.p.m. of this polymer (section 7.3) under 10–25 MPa and at 523 K produced fine (less than 0.1 μm diameter) powders when the supercritical fluid pressure was 24 MPa. However, polycarbosilane fibres 1 μm diameter, 20–160 μm in length were obtained for higher temperatures, typically 648 K. The fluid density drop on expansion promotes nucleation of powders whereas fibre formation probably arose by precipitation, onto the nozzle wall, of solute that was then drawn out to polymer filaments by high-velocity expanding gas.

10.5 Freeze-drying

A recurrent theme for non-conventional powder syntheses is the attainment of greater homogeneity on the molecular scale than is obtainable by powder mixing and hence greater homogeneity in sintered components after fabrication. However, the method used to dry a powder, although not part of the synthesis, can have a profound effect on its pressing and sintering behaviour. Tray-drying was mentioned in section 4.6 in connection with ThO_2 gel fragments but when applied to oxide and hydroxide slurries hard cakes are often produced that need to be crushed and sieved before pressing. Capillary forces whose magnitude are described by the Laplace equation (5.12) and which contribute to pore collapse in alcogels cause a similar effect on tray-dried slurries. This can result in particle agglomerates difficult to break down with subsequent void formation on pressing (Real, 1986). In addition, direct removal of water can enhance particle adhesion through hydrogen bonding.

Spray-drying is used industrially to convert liquid into powder in a one-step process. Liquid or feed, which can be an aqueous oxide slurry, is atomised by using a nozzle or rotating disc and brought into contact with the drying air so that rapid evaporation of water occurs from droplets. The technique has been applied to pharmaceuticals (Nielsen, 1982) and advanced ceramics such as stabilised zirconias (figure 2.1); spray-dried Al_2O_3 powder is shown in figure 4.11. Spray-drying often produces uniform distribution within spheres but the hardness of these particles can lead to voids on sintering (Real, 1986).

Certain drying techniques such as azeotropic distillation (figure 2.1) do not remove water directly and result in reduced agglomeration. Dole & co-workers (1978) showed that a combination of toluene and acetone washes for stripping water from lanthanide, hafnium and zirconium hydroxide precipitates enabled densification of powder compacts to near theoretical density. Dehydration of sol droplets to ThO_2 gel spheres in an alcohol was described earlier (section 4.6) and a similar technique has been combined with the citrate gel process. Mulder (1970) co-precipitated mixed citrate solutions by spraying droplets into an alcohol, and oxides including $MgTiO_3$, Pb_2GdTaO_6, $Zn_3Nb_2O_8$ and $Pb_{0.3}Sr_{0.3}Ba_{0.4}TiO_3$ were produced on decomposition of powders. In a later application (Li, Ni & Yin, 1983), hot-pressed $PbLa_{1-b}(Zr_{0.7}Ti_{0.3})_{1-b/4}O_3$ ($b = 0.08$) derived from alcohol dehydration of citrate solutions was transparent and had better chemical and optical homogeneity than was obtainable by powder mixing techniques.

146

Freeze-drying

In the freeze-drying process, liquid is converted to a solid phase by rapid cooling in, for example, liquid nitrogen, after which solvent is sublimed from solid by heating the latter under reduced pressure. Real (1986) freeze-dried Al_2O_3 slips at below 233 K and showed that the resulting powders consisted of soft agglomerates that could be easily broken down and sintered without void formation. However, when applied to electrolyte solutions rapid cooling retains the homogeneous composition of the solution in freeze-dried salts that are thermally decomposed to oxides. Anderton & Sale (1979) prepared strontium-doped lanthanum cobaltite, for example $La_{0.3}Sr_{0.7}CoO_3$, by freeze-drying nitrate solutions, which were vacuum decomposed at 773 K and then calcined at 1333 K in air. Freeze-drying of nitrate solutions has also been used (Barboux et al., 1987) for obtaining greater compositional homogeneity compared with conventional powder mixing in the high-T_c oxide superconductor $YBa_2Cu_3O_{7-\delta}$ with a particle size less than 5 μm.

Appendix. Determination of particle size

A.1 Introduction

Advanced ceramic powders are usually characterised by measurement of particle size, phase content and impurity levels irrespective of the route used for their synthesis. Particle size is an important parameter for advanced ceramics as it affects the sintering properties and mechanical behaviour of these materials. However, this parameter, in the context of ceramic powders, can refer to primary, aggregated and agglomerated particles as well as a crystallite dimension. A variety of direct and indirect techniques are available for its determination using either dry powders, suspensions or colloidal dispersions. Principal methods are outlined in this appendix.

A.2 Gas adsorption

The BET gas adsorption equation, developed by Brunauer, Emmett & Teller (1938) describes multilayer physical adsorption of gases (usually N_2) onto a solid surface and can be written in the linear form

$$p/v_a(p°-p) = (v_m C_1)^{-1} + (C_1-1)p/p°v_m C_1 \qquad (A.1)$$

where p is the gas pressure, $p°$ the saturation vapour pressure for adsorbate at the temperature of adsorption, v_a the adsorbate volume at relative pressure $p/p°$, v_m the adsorbate volume per unit mass of solid for monolayer coverage and C_1 the BET constant, which is related exponentially to the heat of adsorption in the first adsorbed layer. Surface area, S_A, for a powder is derived by combining v_m with the cross-sectional area of the adsorbate molecule, 0.162 nm^2 for N_2. If the powder consists of spherical particles with radius a and density ρ then

$$a = 3/\rho S_A \qquad (A.2)$$

Gas adsorption

The adsorption isotherm, which is variation of v_a with relative pressure for $0 < p/p° < 1$, contains information on the pore structure of solids. The six types of isotherm are shown in figure A.1 (Gregg, 1986). Pore sizes are divided into three regions. Widths less than 2 nm are characteristic of micropores, those between 2 and 50 nm correspond with mesopores, while macropores have widths greater than 50 nm (Sing *et al.*, 1985). Type I isotherms are obtained with microporous solids while non-porous solids exhibit the reversible Type II isotherms. Type III isotherms are rare and have weak adsorbent–adsorbate interaction whereas Type IV isotherms are associated with capillary condensation in mesoporous solids, giving rise to hysteresis loops. The rare Type V isotherms are, as for Type III isotherms, obtained when the adsorbent–adsorbate interaction is weak whereas Type VI isotherms are for stepwise multilayer adsorption on uniform non-porous surfaces. Pore-size distributions in solids can be calculated by application of the Kelvin equation (9.1) to the complete Type IV isotherm when a model is assumed for pore shape (Sing *et al.*, 1985) and examples of these distributions are shown in figure 5.8.

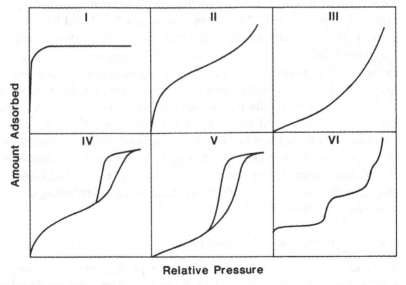

Figure A.1. Types of adsorption isotherm (Gregg, 1986).

Appendix. Determination of particle size

A.3   X-ray line broadening

The diffraction broadening of X-ray peaks obtained from a powder specimen has contributions from the crystallite dimension, D_c, an instrumental broadening and from lattice strain. The crystallite dimension can be obtained from the Scherrer equation (Klug & Alexander, 1974)

$$D_c = K^*\lambda/\theta_{1/2}\cos\theta_B \qquad (A.3)$$

where K^* is a constant (*ca.* 0.9), λ the X-ray wavelength, θ_B the Bragg angle and $\theta_{1/2}$ the pure diffraction broadening of a peak at half-height, that is, broadening due to only the crystallite dimension. Values of D_c do not indicate whether primary, aggregated or agglomerated particles are present in a ceramic powder. If the specimen is poorly crystalline or amorphous to X-rays then this technique is not suitable for determination of particle size.

A.4   Transmission electron microscopy

Although optical microscopy allows direct observation of ceramic powders, its practical limit of resolution is several micrometres. Transmission electron microscopy (TEM) is a suitable technique for measurement of sub-micrometre particle size, and resolution that can be achieved with this method is shown in figures 4.8, 9.5 and 9.7. Transmission electron microscopes consist of five sections (Fryer, 1979). Electrons from a heated cathode, usually tungsten wire, pass through the anode and are brought to a focus by using magnetic lenses onto the specimen, which is contained in an evacuated chamber. Interaction of the incident electron beam with the specimen can give rise to secondary electrons, backscattered electrons, Auger electrons, X-ray emission, cathodoluminescence and inelastic scattering by orbital electrons in the specimen together with elastic scattering by atomic nuclei. Transmitted electrons are amplified by the fourth section in the microscope with a magnification system and finally delivered to an image-recording device such as a fluorescent screen.

A.5   Scanning electron microscopy

In scanning electron microscopy (SEM), a primary electron beam is accelerated by a voltage of 1–50 kV between cathode and anode after

Light scattering

which it is passed through condenser lenses and scanned across the surface of the specimen contained in an evacuated chamber. This electron beam has a diameter 1–10 nm and carries an electron probe current of 10^{-10}–10^{-12} A at the specimen surface (Reimer, 1985). Emissions arise from interaction of the beam and specimen as described for TEM in the previous section and these (usually secondary and back-scattered electrons) are analysed. Scanning electron microscopy is particularly useful for characterisation of coating morphology (figure 5.14) and produces greater three-dimensional detail than TEM for powders (figures 4.11 and 5.16) and fibres (figure 4.13).

A.6 Light scattering

While a beam of light incident on a colloidal dispersion (section 4.2) is transmitted or absorbed, its electric field induces oscillating dipole moments in sol particles, which reradiate or scatter the light in all directions. Colours are observed at different angles to the incident beam when white light is shone through a monodispersed sol and their position depends on particle size as well as refractive index difference between particles and dispersion medium. The intensity ratio for red and green bands, which dominate the colours, varies periodically with angular position (Allen, 1981). This periodicity is characteristic of higher-order Tyndall spectra, which were mentioned earlier in connection with particle size measurements in sulphur sols (section 5.7).

When light with wavelength λ is incident on an isolated spherical particle, radius a (a ca. $\lambda/10$), the Rayleigh scattered intensity, I_R, is given by the expression (Pusey, 1982)

$$I_R = 16\pi^4 a^6 (n_p^2 - n_m^2)^2 / l_0^2 \lambda^4 (n_p^2 + 2n_m^2)^2 \qquad (A.4)$$

where n_p is the refractive index of the particle, n_m the refractive index of the surrounding medium and l_0 the distance from sample to detector. Interference between scattered fields from different regions of the particle occurs as the radius increases and when $a \approx \lambda$ the Rayleigh–Gans scattered intensity, I_{RG}, can be written in the form

$$I_{RG} = I_R P(\phi) \qquad (A.5)$$

where $P(\phi)$ is a shape factor and ϕ is the angle between incident and scattered radiation; $P(0) = 1$ and $P(\phi) < 1$ when $\phi > 0$. For uniform spheres

Appendix. Determination of particle size

$$P(\phi) = [3(\sin Qa - Qa\cos Qa)/(Qa)^3]^2 \qquad (A.6)$$

where Q, the scattering vector is defined as

$$Q = 4\pi\lambda^{-1}\sin(\phi/2) \qquad (A.7)$$

Mie theory applies when $a \gg \lambda$ and results in a complex angular dependence of scattered intensity. Equations (A.4) and (A.5) are the basis of time-averaged light scattering for particle-size determination in colloidal dispersions. The technique has been described in detail by Kerker (1969) but it should be stressed that the size parameter from intensity measurements is the radius of gyration, which, for spheres, has a magnitude $(3/5)^{1/2}a$.

The Rayleigh theory of light scattering used for deriving equation (A.4) is unsatisfactory in condensed media because of interference between scattered field from different particles (Weiner, 1984). Einstein showed that thermal fluctuations cause variation in properties (e.g. density, entropy, concentration) of a pure liquid and were responsible for light scattering. Time-averaged measurements are concerned with spatial variation of these fluctuations but temporal fluctuations produce frequency broadening in the scattered light and contain information on particle motion. Particle size can be measured by using coherent laser light sources with time-dependent light scattering, also called dynamic light scattering, but frequently referred to as photon correlation spectroscopy (PCS); at the time of writing PCS is the principal technique for size determination in colloidal dispersions.

In PCS, the autocorrelation function, $\langle I(0)I(\tau)\rangle$, defined as

$$\langle I(0)I(\tau)\rangle = \lim_{t_e \to \infty} t_e^{-1} \int_0^{t_e} I(t)I(t-\tau)\mathrm{d}t \qquad (A.8)$$

is measured where I is the scattered intensity at time t, τ and $t-\tau$, and t_e is the duration time for the experiment. The diffusion coefficient, D, for particles in a dispersion can be obtained from equation (A.8) and related to sphere radius, a, through the Stokes–Einstein equation

$$D = kT/6\pi\eta a \qquad (A.9)$$

where η is the viscosity of the medium and other variables have been previously defined. The radius in equation (A.9) is, strictly, a hydrodynamic value and may include contributions from adsorbed or solvated layers. Berne & Pecora (1976) and Chu (1974) have given detailed mathematical descriptions of PCS, while instructive accounts have been described by Pusey (1982) and more recently by Weiner (1984).

A.7   Small-angle X-ray and neutron scattering

In small-angle X-ray scattering (SAXS) the intensity I of coherent elastically scattered X-rays is measured as a function of Q (equation (A.7)) for ϕ values less than about 0.05 rad. Unlike diffraction at large Bragg angles, which contains information on crystal structure and crystallite size (section A.3), SAXS detects inhomogeneities in electron density in the specimen. These inhomogeneities on the colloidal scale can refer to particle size in a dry powder or dispersion and this technique has been widely applied to macromolecules. Scattered intensity is expressed by the Guinier equation (Porod, 1982)

$$I = I_0 \exp \left(-Q^2 r_g^2/3 \right) \tag{A.10}$$

where Q is the scattering vector defined in equation (A.7), I_0 the value of I at zero Q and r_g is the radius of gyration for powder particles, which may be calculated from the linear variation of $\ln I$ with Q^2; Guinier & Fournet (1955) have given a detailed mathematical description of SAXS.

Small-angle neutron scattering (SANS) is a technique related to SAXS in that the Guinier equation (A.10) can be applied to the scattering of neutrons. Whereas X-ray scattering arises from interaction between X-rays and orbital electrons, a neutron beam (wavelength 0.1–2 nm) is coherently scattered in all directions from atomic nuclei whose neutron scattering length affect this process (Ottewill, 1982). Neutrons have greater penetrating power than light and SANS is particularly suited to particle-size determination in concentrated opaque dispersions, while light scattering is appropriate for dilute systems. Scattering lengths do not vary uniformly with atomic number so that while deuterium scatters coherently, hydrogen is an incoherent scatterer. This variation in scattering length makes SANS a powerful tool for investigating fine structure in colloidal systems, for example adsorbed layers on particles, as the contrast between regions of the particle and its background can be varied until neutrons 'see' only parts of the scatterer. Bacon (1977) has described the application of neutron scattering, including SANS, to chemical systems in more detail.

Appendix. Determination of particle size

A.8   Sedimentation methods

Size determination by sedimentation methods involves settling of a suspension or colloidal dispersion either under gravity or, for small particles with diameters less than about 3 μm, in a centrifugal field, which overcomes the influence of Brownian motion on this size range. When a sphere with radius a falls through a liquid under laminar flow conditions the viscous drag can be equated to effective particle weight by the Stokes equation (Society for Analytical Chemistry, 1968)

$$a^2 = 9\eta l_1 (\rho_p - \rho_l)^{-1}/2Gt \qquad (A.11)$$

where ρ_p is the particle density, ρ_l the liquid density, η the medium viscosity, G the gravitational constant and l_1 the sedimentation distance at time t. The Stokes equation is modified in the presence of a centrifugal field (Veale, 1972), hence

$$a^2 = 9\eta(\rho_p - \rho_l)^{-1}\ln(l_2/l_3)/2 \; \Omega^2 t \qquad (A.12)$$

where Ω is the angular rotational velocity, l_3 the radius of rotation for the particle at zero time, and l_2 the distance between particle and axis of rotation at time t. Equations (A.11) and (A.12) are the basis of sedimentation methods for particle-size analysis.

A common feature of sedimentation techniques is that particle-size distributions can be obtained from them whereas light scattering, SAXS and SANS produce average sizes. For example, in the Andreassen Pipette (Rideal, 1982; Lloyd & Treasure, 1976) a powder suspension is allowed to settle under gravity and sample volumes are withdrawn at pre-determined times so that the particle radius calculated from equation (A.11) can be equated with a solids fraction obtained by drying the sample. Removal of suspension during settling is avoided by monitoring the sedimentation process with optical or X-ray transmittance and although gravity settling is applicable to sub-micrometre sizes it is particularly useful for diameters above 5 μm.

Enhanced settling rates in a centrifugal field extend the size range down to about 0.1 μm by using, for example, the Ladal centrifuge, which has been discussed by Allen (1981) who has also given mathematical descriptions of sedimentation processes. Although ultracentrifugation (Bowen, 1970) is used for molecular-weight determination as well as measurement of diffusion coefficients for biological macromolecules and polymers, the latter parameter is nowadays more often obtained by photon correlation spectroscopy (section A.6). The choice of sedimentation technique depends not only on size range but also the cost of equipment, and Rideal (1982) has compared the latter for both gravitational and centrifugal methods.

A.9   Chromatographic methods

Chromatographic methods are concerned with separation of components in a mixture by utilising affinities of different solute molecules for stationary and mobile phases and include high-performance liquid chromatography (Knox, 1976) when the mobile phase is subjected to a high pressure. One technique that has been applied for characterisation of ceramic precursors such as polycarbosilanes (section 7.3) is gel-permeation chromatography (GPC) in which molecules are separated according to their size. A mixture of polymers with varying molecular weight and hence size is injected onto the top of a column packed with the stationary phase, a porous oxide powder. Molecules that are too big to enter pores in this powder are eluted first by the mobile phase whereas small molecules, which permeate through the pores, are the last to be eluted (Bristow, 1976) so that molecular-weight distributions can be obtained.

A.10   Sieving

Sieving is particularly suited for powders with particle size greater than 56 μm (Lloyd & Treasure, 1976) but can be used down to 5 μm with electroformed micromesh sieves. Separation can be carried out manually or by machine and a related techique used in industry is classification. Allen (1981) has described these techniques in more detail and his book contains a comprehensive account of many methods used for particle size determination.

References

Chapter 1

AE Development Limited. (1985). Engineering ceramics. In *Official Reference Book and Buyers' Guide 1985/6*, pp. 430–2. London: Institution of Engineering Designers.

American Ceramic Society Incorporated. (1987). *Ceramic Source '88*, vol. 3. Ohio: American Ceramic Society.

Bednorz, J. G. & Müller, K. A. (1986). Possible high-T_c superconductivity in the Ba–La–Cu–O system. *Zeitschrift für Physik B*, **64**, 189–93.

Bednorz, J. G., Takashige, M. & Müller, K. A. (1987). Susceptibility measurements support high-T_c superconductivity in the Ba–La–Cu–O system. *Europhysics Letters*, **3**, 379–85.

Bell, A. J. (1987). Materials and processes for improved electroceramics. In *Fine Chemicals for the Electronics Industry*, ed. P. Bamfield, pp. 176–93. London: Royal Society of Chemistry.

Boretos, J. W. (1987). Advances in bioceramics. *Advanced Ceramic Materials*, **2**, 15–22, 30.

Briscoe, E. M. (1986). The use of advanced ceramics in engines. *Chemistry and Industry*, 641–5.

Cantagrel, M. (1986). Development of ceramics for electronics. *American Ceramic Society Bulletin*, **65**, 1248–9.

Charles H. Kline and Company. (1987). Demand for ceramic coatings. *Ceramic Industries Journal*, April 1987, 18–20.

Clarke, D. R. (1987). The development of high-T_c ceramic superconductors. An introduction. *Advanced Ceramic Materials*, **2**, special issue number 3B, 273–92.

Frederikse, H. P. R., Schooley, J. F., Thurber, W. R., Pfeiffer, E. & Hosler, W. R. (1966). Superconductivity in ceramic mixed titanates. *Physical Review Letters*, **16**, 579–81.

Gavaler, J. R. (1973). Superconductivity in Nb–Ge films above 22 K. *Applied Physics Letters*, **23**, 480–2.

Heinrich, J. (1985). Contribution to the technology of hot-isostatic pressing of some non-oxide ceramics. *Ceramic Forum International*, **62**, 222–8.

Jack, D. H. (1986). Ceramic tool materials. In *Applications of Engineering Ceramics. Non IC Engine Applications*, pp. 11–16. London: The Institution of Mechanical Engineers.

References

Jack, K. H. (1986). Sialons. A study in materials development. In *Non-Oxide Technical and Engineering Ceramics*, ed. S. Hampshire, pp. 1–28. London: Elsevier Applied Science.

Jin, S., Sherwood, R. C., van Dover, R. B., Tiefel, T. H. & Johnson, Jr, D. W. (1987). Fabrication of 91 K superconducting coils. In *Extended Abstracts of Symposium S, High Temperature Superconductors*, vol. EA–11, ed. D. U. Gubser & M. Schluter, pp. 219–20, Anaheim, California, 23–24 April 1987. Organised by the Materials Research Society.

Johnston, D. C., Prakash, H., Zachariasen, W. H. & Viswanathan, R. (1973). High temperature superconductivity in the Li–Ti–O ternary system. *Materials Research Bulletin*, **8**, 777–84.

Knoch, H. (1989). Non-oxide technical ceramics. In *2nd European Symposium on Engineering Ceramics*, ed. F.L. Riley, pp 151–69. London: Elsevier Applied Science.

Kulwicki, B. M. (1984). Ceramic sensors and transducers. *Journal of Physics and Chemistry of Solids*, **45**, 1015–31.

Lackey, W. J., Stinton, D. P., Cerny, G. A., Schaffhauser, A. C. & Fehrenbacher, L. L. (1987). Ceramic coatings for advanced heat engines. A review and projection. *Advanced Ceramic Materials*, **2**, 24–30.

Lay, L. A. (1983). *Corrosion Resistance of Technical Ceramics*. London: Her Majesty's Stationery Office.

Meetham, G. W. (1986). Advanced ceramics in aerospace. *Chemistry and Industry*, pp. 636–40.

Michel, C. & Raveau, B. (1984). Oxygen intercalation in mixed valence copper oxides related to the perovskites. *Revue de Chimie Minérale*, **21**, 407–25.

Morrell, R. (1985). *Handbook of Properties of Technical & Engineering Ceramics. Part I. An Introduction for the Engineer and Designer*. London: Her Majesty's Stationery Office.

Norton, F. H. (1968). *Refractories*, 4th edition. New York: McGraw-Hill Book Company.

Onnes, H. K. (1911a). Further experiments with liquid helium. D. On the change of the electrical resistance of pure metals at very low temperatures. V. The disappearance of the resistance of mercury. *Konink Akad van Wetenschappen (Amsterdam)*, **14**, 113–5.

Onnes, H. K. (1911b). Further experiments with liquid helium. G. On the electrical resistance of pure metals. VI. On the sudden change in the rate at which the resistance of mercury disappears. *Konink Akad van Wetenschappen (Amsterdam)*, **14**, 818–21.

Raveau, B. & Michel, C. (1987). Crystal chemistry of copper oxides related to the perovskite relationships with superconductivity. *Chemtronics*, **2**, 107–15.

Shaw, K. (1972). *Refractories and Their Uses*. London: Applied Science Publishers.

Sleight, A. W., Gillson, J. L. & Bierstedt, P. E. (1975). High temperature superconductivity in the $BaPb_{1-b}Bi_bO_3$ system. *Solid State Communications*, **17**, 27–8.

Taguchi, M. (1987). Applications of high-technology ceramics in Japanese automobiles. *Advanced Ceramic Materials*, **2**, 754–62.

References

Takagi, H., Uchida, S., Kitazawa, K. & Tanaka, S. (1987). High T_c superconductivity of La–Ba–Cu oxides. II. Specification of the superconducting phase. *Japanese Journal of Applied Physics*, **26**, L123–4.

Tummala, R. R. & Shaw, R. B. (1987). Ceramics in microelectronics. *Ceramics International*, **13**, 1–11.

Tuttle, B. A. (1987). Electronic ceramic thin films: trends in research and development. *Materials Research Society Bulletin*, **12**, 40–5.

Wen, L. S., Qian, S. W., Hu, Q. Y., Yu, B. H., Zhao, H. W., Guan, K., Fu, L. S. & Yang, Q. Q. (1987). Plasma-sprayed high T_c superconductor coatings. *Thin Solid Films*, **152**, L143–5.

Wu, M. K., Ashburn, J. R., Torng, C. J., Hor, P. H., Meng, R. L., Gao, L., Huang, Z. J., Wang, Y. Q. & Chu, C. W. (1987). Superconductivity at 93 K in a new mixed phase Y–Ba–Cu–O compound system at ambient pressure. *Physical Review Letters*, **58**, 908–10.

Chapter 2

Cable, M. (1984). Principles of glass-making. In *Glass: Science and Technology*, ed. D. R. Uhlmann & N. J. Kreidl, pp. 1–14, London: Academic Press, Inc.

Cava, R. J., Batlogg, B., van Dover, R. B., Murphy, D. W., Sunshine, S., Siegrist, T., Remeika, J. P., Rietman, E. A., Zahurak, S. & Espinosa, G. P. (1987). Bulk superconductivity at 91K in single-phase oxygen-deficient perovskite $YBa_2Cu_3O_{9-\delta}$. *Physical Review Letters*, **58**, 1676–9.

Clough, D. J. (1985). ZrO_2 powders for advanced and engineered ceramics. *Ceramic Engineering and Science Proceedings*, **6**, 1244–60.

Doyle, P. J. (1979). *Glass-Making Technology*. Redhill: Portcullis Press.

Evans, K. A. & Brown, N. (1981). Speciality inorganic aluminium compounds. In *Speciality Inorganic Chemicals*, ed. R. Thompson, pp. 164–95. London: Royal Society of Chemistry.

Jorgensen, J. D., Schüttler, H. B., Hinks, D. G., Capone, II, D. W., Zhang, K., Brodsky, M. B. & Scalapino, D. J. (1987). Lattice instability of high-T_c superconductivity in $La_{2-b}Ba_bCuO_4$. *Physical Review Letters*, **58**, 1024–7.

MacZura, G., Carbone, T. J. & Hart, L. D. (1987). Annual ceramic mineral resources review: alumina. *American Ceramic Society Bulletin*, **66**, 753–4.

Philips Electrical Industries Limited. (1954). Improvements in or relating to semi-conductive material. *United Kingdom Patent* 714,965.

Stookey, S. D. (1964). Low expansion glass-ceramic and method of making it. *United States Patent* 3,157,522.

Toya Soda Manufacturing Company. (1984). *TSK Ceramics. Technical Bulletin, number Z-051. The TZ-3Y Process*.

Wilkinson, W. L. (1981). Chemistry of the nuclear fuel cycle. In *Energy and Chemistry*, ed. R. Thompson, pp. 164–86. London: Royal Society of Chemistry.

Yang, Z. Z., Yamada, H. & Miller, G. R. (1985). Synthesis and characterisation of high-purity $CaTiO_3$. *American Ceramic Society Bulletin*, **64**, 1550–4.

References

Chapter 3

Alford, N. McN., Birchall, J. D. & Kendall, K. (1987). High-strength ceramics through colloidal control to remove defects. *Nature*, **330**, 51–3.

Davidge, R. W. (1979). *Mechanical Behaviour of Ceramics*, Cambridge: Cambridge University Press.

Griffith, A. A. (1920). The phenomena of rupture and flow in solids. *Philosophical Transactions of The Royal Society of London*, A**221**, 163–98.

Holloway, D. G. (1986). The fracture behaviour of glass. *Glass Technology*, **27**, 120–33.

Hull, D. (1985). *An Introduction to Composite Materials*. Cambridge: Cambridge University Press.

James, P. J. (1983). Principles of isostatic pressing. In *Isostatic Pressing Technology*, ed. P. J. James, pp. 1–27. London: Applied Science Publishers.

Kingery, W. D. (1983). Powder preparation. In *Materials Science Monographs. Ceramic Powders: Preparation, Consolidation and Sintering*, vol. 16, ed. P. Vincenzini, pp. 3–18, Amsterdam: Elsevier.

Kingery, W. D., Bowen, H. K. & Uhlmann, D. R. (1976). *Introduction to Ceramics, 2nd Edition*. New York: John Wiley & Sons.

Lange, F. F. (1984). Structural ceramics: A question of fabrication reliability. *Journal of Materials for Energy Systems*, September 1984, 107–13.

Mangels, J. A. & Tennenhouse, G. J. (1980). Densification of reaction-bonded silicon nitride. *American Ceramic Society Bulletin*, **59**, 1216–18.

Morrell, R. (1985). *Handbook of Properties of Technical & Engineering Ceramics. Part I. An Introduction for the Engineer and Designer*. London: Her Majesty's Stationery Office.

Newkirk, M. S., Lesher, H. D., White, D. R., Kennedy, C. R., Urquhart, A. W. & Claar, T. D. (1987). Preparation of Lanxide ceramic matrix composites. Matrix formation by the directed oxidation of molten metals. *Ceramic Engineering and Science Proceedings*, **8**, 879–85.

Petzow, G. & Kaysser, W. A. (1984). Basic mechanisms of liquid phase sintering. In *Materials Science Monographs: Sintered Metal–Ceramic Composites*, vol. 25, ed. G. S. Upadhyaya, pp. 51–70. Amsterdam: Elsevier.

Phelps, G. W. & McLaren, M. G. (1978). Particle size distribution and slip properties. In *Ceramic Processing Before Firing*, ed. G. Y. Onoda, Jr & L. L. Hench, pp. 211–25. New York: John Wiley & Sons.

Popper, P. (1983). Sintering of silicon nitride, a review. In *Progress in Nitrogen Ceramics*, ed. F. L. Riley, pp. 187–210. The Hague: Martinus Nijhoff.

Richerson, D. W. (1982). *Modern Ceramic Engineering. Properties, Processing and Use in Design*. New York: Marcel Dekker, Inc.

Richter, D., Haour, G. & Richon, D. (1985). Hot isostatic pressing. *Materials and Design*, **6**, 303–5.

Thomas, W. J. (1987). The catalytic monolith. *Chemistry and Industry*, pp. 315–9.

Thurnauer, H. (1958). Controls required and problems encountered in production dry pressing. In *Ceramic Fabrication Processes*, ed. W. D. Kingery, pp.

References

62–70. USA: John Wiley & Sons, Chapman Hall and The Technology Press of The Massachusetts Institute of Technology.

Wiederhorn, S. M. (1984). Brittle fracture and toughening mechanisms in ceramics. *Annual Review of Materials Science*, **14**, 373–403.

Chapter 4

Arnold, E. D. (1962). Radiation hazards for recycled ^{233}U–thorium fuels. In *Proceedings of the Thorium Fuel Cycle Symposium*, Gatlinburg, Tennessee, 5–7 December, TID 7650, book 1, pp. 253–284.

Baes, Jr, C. F. & Mesmer, R. E. (1976). *The Hydrolysis of Cations*. New York: John Wiley & Sons.

Birchall, J. D. (1983). The preparation and properties of polycrystalline aluminium oxide fibres. *Transactions of the Journal of the British Ceramic Society*, **82**, 143–5.

Bones, R. J. & Woodhead, J. L. (1974). Improvements in or relating to particulate indium oxide. *United Kingdom Patent* 1,351,113.

Bonner, F. J., Kordas, G. & Kinser, D. L. (1985). Sol–gel glasses by non-aqueous processes. *Journal of Non-Crystalline Solids*, **71**, 361–71.

Brambilla, G., Gerontopoulos, P. & Neri, D. (1970). The SNAM process for the preparation of ceramic nuclear fuel microspheres: Laboratory studies. *Energia Nucleare*, **17**, 217–24.

Cairns, J. A., Segal, D. L. & Woodhead, J. L. (1984). The role of gel processing in the preparation of catalyst supports. *Materials Research Society Symposium Proceedings*, **32**, 135–8.

Chandrashekhar, G. V. & Shafer, M. W. (1986). Dielectric properties of sol–gel glasses. *Material Research Society Symposium Proceedings*, **73**, 705–8.

Cogliatti, G., DeLeone, R., Guidotti, G. R., Lanz, R., Lorenzini, L., Mezi, E. & Scibona, G. (1964). The preparation of dense particles of thorium and uranium oxides. In *Third United Nations International Conference on the Peaceful Uses of Atomic Energy*, A/CONF.28/P/555.

Dean, O. C., Haws, C. C., Kleinsteuber, A. T. & Snider, J. W. (1962). The sol-gel process for preparation of thoria based fuels. In *Proceedings of the Thorium Fuel Cycle Symposium*, Gatlinburg, Tennessee, 5–7 December, TID 7650, book II, pp. 519–42.

Faraday, M. (1857). On the experimental relations of gold (and other metals) to light. *Philosophical Transactions of the Royal Society*, **147**, 145.

Ferguson, D. E., Dean, O. C. & Douglas, D. A. (1964). The sol–gel process for the remote preparation and fabrication of recycle fuels. In *Third United Nations International Conference on the Peaceful Uses of Atomic Energy*, A/CONF.28/P/237.

Flory, P. J. (1974). Introductory lecture. *Discussions of the Faraday Society*, **57**, 7–18.

Ganguly, C., Langen, H., Zimme, E. & Merz, E. R. (1986). Sol–gel microsphere pelletization process for fabrication of high density ThO_2–2%UO_2 fuel

160

for advanced pressurized heavy water reactors. *Nuclear Technology*, **73**, 84–95.

Graham, T. (1861). Liquid diffusion applied to analysis. *Philosophical Transactions of the Royal Society*, **151**, 183–224.

Griffiths, J. (1986). Synthetic mineral fibres: From rocks to riches. *Industrial Minerals*, **228**, 20–43.

Haas, P. A. & Clinton, S. D. (1966). Preparation of thoria and mixed-oxide microspheres. *Industrial & Engineering Chemistry Product Research and Development*, **5**, 236–44.

Hamaker, H. C. (1937). The London–van der Waals attractions between spherical particles. *Physica*, **4**, 1058–72.

Hardy, C. J. (1968). Status report from the United Kingdom; Sol–gel and gel-precipitation processes. In *Sol–Gel Processes for Ceramic Nuclear Fuels*, pp. 33–42. Vienna: International Atomic Energy Agency.

Hardy, C. J., Buxton, S. R. & Lloyd, M. H. (1967a). *Preparation of Lanthanide Oxide Microspheres by Sol–Gel Methods*. Oak Ridge National Laboratory Report ORNL-4000. Tennessee: Oak Ridge National Laboratory.

Hardy, C. J., Buxton, S. R. & Willmarth, T. E. (1967b). Chemical and electron optical studies of sols, gels and oxide microspheres prepared from rare-earth hydroxides. *Proceedings of the 6th Rare Earth Research Conference*, Oak Ridge, Tennessee, 3–5 May 1967, pp. 175–86.

Hermans, M. E. A. (1968). Status report from the Netherlands; Sol–gel fuel studies. In *Sol–Gel Processes for Ceramic Nuclear Fuels*, pp. 21–32. Vienna: International Atomic Energy Agency.

Hermans, M. E. A. & Slooten, H. S. G. (1964). Preparation of UO_2 and ThO_2 powder in the subsieve range. In *Third United Nations International Conference on the Peaceful Uses of Atomic Energy*, A/CONF.28/P/634.

Hiemenz, P. C. (1986). *Principles of Colloid and Surface Chemistry*, 2nd edition, New York: Marcel Dekker Inc.

Imperial Chemical Industries Limited, (1975). *Saffil Inorganic Fibres: Alumina Fibres*. Technical Leaflet NV/3369/1Ed/42/1075.

Kelly, J. L., Kleinsteuber, A. T., Clinton, S. D. & Dean, O. C. (1965). Sol–gel process for preparing spheroidal particles of the dicarbides of thorium and thorium–uranium mixtures. *Industrial & Engineering Chemistry Process Design and Development*, **4**, 212–16.

Kepert, D. L. (1972). *The Early Transition Metals*. London: Academic Press.

Lauf, R. J. & Bond, W. D. (1984). Fabrication of high-field zinc oxide varistors by sol–gel processing. *Bulletin American Ceramic Society*, **63**, 278–81.

Leitheiser, M. A. & Sowman, H. G. (1982). Non-fused aluminium oxide-based abrasive material. *United States Patent* 4,314,827.

Livage, J. & Lemerle, J. (1982). Transition metal oxide gels and colloids. *Annual Reviews of Materials Science*, **12**, 103–22.

Lyklema, J. (1985). Interfacial electrochemistry of disperse systems. *The Journal of Materials Education*, **7**, 205–38.

Mahanty, J. & Ninham, B. W. (1976). *Dispersion Forces*. New York: Academic Press.

References

Marples, J. A. C., Nelson, R. L., Potter, P. E. & Roberts, L. E. J. (1981). Chemistry in the development of nuclear power. In *Energy and Chemistry*, ed. R. Thompson, pp. 131–63. London: Royal Society of Chemistry.

Morganite Ceramic Fibres Limited. (1981). *Triton Kaowool Ceramic Fibre: Product Data*. Technical Leaflet MCF 39/1.

Morton, M. J., Birchall, J. D. & Cassidy, J. E. (1974). Fibres. *United Kingdom Patent* 1,360,197.

Napper, D. H. (1982). Polymeric stabilisation. In *Colloidal Dispersions*, 1st edition, ed. J. W. Goodwin, pp. 99–128. London: Royal Society of Chemistry.

Nelson, R. L., Ramsay, J. D. F., Woodhead, J. L., Cairns, J. A. & Crossley, J. A. A. (1981). The coating of metals with ceramic oxides via colloidal intermediates. *Thin Solid Films*, **81**, 329–37.

Overbeek, J. Th. G. (1982). How colloid stability affects the behaviour of suspensions. *Materials Science Research*, **17**, 25–43.

Rabinovich, E. M., Johnson, D. W., MacChesney, J. B. & Vogel, E. M. (1982). Preparation of transparent high-silica glass articles from colloidal gels. *Journal of Non-Crystalline Solids*, **47**, 435–9.

Rabinovich, E. M., MacChesney, J. B., Johnson, D. W., Stimpson, J. R., Meagher, B. W., Dimarcelto, F. V., Wood, D. L. & Sigety, E. A. (1984). Sol–gel preparation of transparent silica glass. *Journal of Non-Crystalline Solids*, **63**, 155–61.

Ramsay, J. D. F. (1978). Porous silica supports for high performance exclusion chromatography. In *Chromatography of Synthetic and Biological Polymers*, vol. 1, ed. R. Epton, pp. 339–43. Chichester: Ellis Horwood.

Ramsay, J. D. F. (1986). Recent developments in the characterisation of oxide sols using small angle neutron scattering techniques. *Chemical Society Reviews*, **15**, 335–71.

Reeve, K. D. & Ringwood, A. E. (1983). The Synroc process for immobilising high-level nuclear wastes. In *Proceedings of an International Conference on Radioactive Waste Management*, Seattle, 16–20 May 1983, vol. 2, pp. 307–24. Vienna: International Atomic Energy Agency.

Roy, R. (1969). Gel route to homogeneous glass preparation. *Journal of the American Chemical Society*, **52**, 344.

Scherer, G. W. & Luong, J. C. (1984). Glasses from colloids. *Journal of Non-Crystalline Solids*, **63**, 163–72.

Scott, K. T. & Cross, A. G. (1986). Near-net shape fabrication by thermal spraying. *Proceedings of the British Ceramic Society*, **38**, 203–11.

Scott, K. T. & Woodhead, J. L. (1982). Gel processed powders for plasma spraying. *Thin Solid Films*, **95**, 219–25.

Segal, D. L. & Woodhead, J. L. (1986). New developments in gel processing. *Proceedings of the British Ceramic Society*, **38**, 245–50.

Selmi, F. (1845). Studies on the demulsion of silver chloride. *Nuovi Annali delle Scienze Naturali di Bologna*, series II, **4**, 146.

Selmi, F.(1847). A study of the pseudo-solutions of Prussian blue and of the influence of salts in destroying them. *Nuovi Annali delle Scienze Naturali di Bologna*, series II, **8**, 401.

References

Sowman, H. G. & Johnson, D. D. (1985). Ceramic oxide fibres. *Ceramic & Engineering Science Proceedings*, **6**, 1221–30.

Szweda, A., Hendry, A. & Jack, K. H. (1981). The preparation of silicon nitride from silica by sol–gel processing. *Proceedings of the British Ceramic Society*, **31**, 107–18.

Tabor, D. (1982). Attractive surface forces. In *Colloidal Dispersions*, 1st edn., ed. J. W. Goodwin, pp. 23–46. London: Royal Society of Chemistry.

Turner, C. W.(1986). *Colloid Chemistry and its Application to the Production of Recycled Fuels by Sol–Gel Processes*. Atomic Energy of Canada Limited Report AECL-8062.

Verwey, E. J. W. & Overbeek, J. Th. G. (1948). *Theory of the Stability of Lyophobic Colloids*. Amsterdam: Elsevier.

Vold, M. J. (1961). The effect of adsorption on the van der Waals interaction of spherical colloidal particles. *Journal of Colloid Science*, **16**, 1–12.

Woodhead, J. L. (1970). Improvements in or relating to zirconium compounds. *United Kingdom Patent* 1,181,794.

Woodhead, J. L. (1974). Improvements in or relating to the production of ceria. *United Kingdom Patent* 1,342,893.

Woodhead, J. L. (1975). Improvements in or relating to titanium dioxide gels and sols. *United Kingdom Patent* 1,412,937.

Woodhead, J. L. (1984). Sol–gel processes to ceramic particles using inorganic precursors. *Journal of Materials Education*, **6**, 887–926.

Woodhead, J. L. & Segal, D. L. (1984). Sol–gel processing. *Chemistry in Britain*, **20**, 310–3.

Woodhead, J. L. & Segal, D. L. (1985). Sol–gel processes for the preparation of electrically conducting ceramic powders. *Proceedings of the British Ceramic Society*, **36**, 123–8.

Yarborough, W. A., Gurujava, T. R. & Cross, L. E. (1987). Materials for IC packaging with very low permittivity via colloidal sol–gel processing. *American Ceramic Society Bulletin*, **66**, 692–8.

Chapter 5

Arfsten, N. F., Kaufmann, R. & Dislich, H. (1984). Sol–gel derived indium tin oxide coatings. In *Ultrastructure Processing of Ceramics, Glasses and Composites*, ed. L. L. Hench & D. R. Ulrich, pp. 189–96. New York: John Wiley & Sons.

Artaki, I., Sinha, S., Irwin, A. D. & Jones, J. (1985). ^{29}Si NMR study of the initial stage of the sol–gel process under high pressure. *Journal of Non-Crystalline Solids*, **72**, 391–402.

Ashley, C. S. & Reed, S. T. (1984). *Sol–Gel Derived AR Coatings for Solar Receivers*. Sandia National Laboratories Report SAND-84-0662, also DE 85000192. Albuquerque: Department of Energy, USA.

Barringer, E. A. & Bowen, H. K. (1985a). High-purity monodispersed TiO_2 powder by hydrolysis of titanium tetraethoxide. I Synthesis and physical properties. *Langmuir*, **1**, 414–20.

References

Barringer, E. A. & Bowen, H. K. (1985b). Ceramic powder processing. *Ceramic Engineering and Science Proceedings*, **5**, 285–97.

Barringer, E. A. & Bowen, H. K. (1982). Formation, packing and sintering of monodispersed TiO_2 powders. *Journal of the American Ceramic Society*, **65**, C199–201.

Barringer, E. A., Jubb, N., Fegley, B., Pober, R. L. & Bowen, H. K. (1984). Processing monosized powders. In *Ultrastructure Processing of Ceramics, Glasses and Composites*, ed. L. L. Hench & D. R. Ulrich, pp. 315–33, New York: John Wiley & Sons.

Blum, J. B. & Ryan, J. W. (1986). Gas chromatography study of the acid catalysed hydrolysis of tetraethylorthosilicate. *Journal of Non-Crystalline Solids*, **81**, 221–6.

Bradley, D. C., Chakravarti, B. N., Chatterjee, A. K., Wardlaw, W. & Whiley, A. (1958). Niobium and tantalum mixed alkoxides. *Journal of the Chemical Society*, pp. 99–101.

Bradley, D. C., Harder, B. & Hudswell, F. (1957). Plutonium alkoxides. *Journal of the Chemical Society*, p. 3318.

Bradley, D. C., Mehrotra, R. C. & Gaur, D. P. (1978). *Metal Alkoxides*. London: Academic Press.

Bradley, D. C., Mehrotra, R. C. & Wardlaw, W. (1952). Structural chemistry of the alkoxides. Part III. Secondary alkoxides of silicon, titanium and zirconium. *Journal of the Chemical Society*, pp. 5020–3.

Bradley, D. C., Saad, M. A. & Wardlaw, W. (1954). Alcoholates of thorium tetrachloride. *Journal of the Chemical Society*, pp. 2002–5.

Bradley, D. C., Wardlaw, W. & Whitley, A. (1956). Structural chemistry of the alkoxides. Part V. Isomeric butoxides and pentyloxides of quinquevalent tantalum. *Journal of the Chemical Society*, pp. 1139–42.

Brinker, C. J. (1982). Formation of oxynitride glasses by ammonolysis of gels. *Journal of the American Ceramic Society*, **65**, C4–5.

Brinker, C. J. & Harrington, M. S. (1981). Sol–gel derived antireflective coatings for silicon. *Solar Energy Materials*, **5**, 159–72.

Brinker, C. J., Keefer, K. D., Schaefer, D. W., Assink, R. A., Kay, B. D. & Ashley, C. S. (1984). Sol–gel transition in simple silicates. *Journal of Non-Crystalline Solids*, **63**, 45–59.

Brinker, C. J., Roth, E. P., Scherer, G. W. & Tallant, D. R. (1985). Structural evolution during the gel to glass conversion. *Journal of Non-Crystalline Solids*, **71**, 171–85.

Brow, R. K. & Pantano, C. G.(1986). Oxidation resistant sol–gel derived silicon oxynitride thin films. *Applied Physics Letters*, **48**, 27–9.

Bruce, R. W. & Kordas, G. (1986). Iron oxide and yttrium iron oxide thin films. A test of the feasibility of producing and measuring micro- and millimetre-wave materials. *Materials Research Society Symposium Proceedings*, **73**, 679–84.

Budd, K. D., Dey, S. K. & Payne, D. A. (1985). Sol–gel processing of $PbTiO_3$, $PbZrO_3$, PZT and PLZT thin films. *Proceedings of the British Ceramic Society*, **36**, 107–21.

References

Condea Chemie GmBH. (1986). *Condea Chemie: Information about the Company*. Technical leaflet SH424-639A, July 1986.

Congshen, Z., Lisong, H., Fuxi, G. & Zhonghong, J. (1984). Low temperature synthesis of ZrO_2–TiO_2–SiO_2 glasses from $Zr(NO_3)_4.5H_2O$, $Si(OC_2H_5)_4$ and $Ti(OC_4H_9)_4$ by the sol–gel method. *Journal of Non-Crystalline Solids*, **63**, 105–15.

Defay, R., Prigogine, I., Bellemans, A. & Everett, D. A. (1966). *Surface Tension and Adsorption*. London: Longman.

Degussa Company. (1981). *Technical Bulletin Pigments No. 6: Hydrophobic Aerosil, Manufacture, Properties and Applications*, March 1981.

Dislich, H. (1985a). Sol–gel derived dip coatings. *Vide Couches Minces*, **40**, 261–8.

Dislich, H. (1985b). Sol–gel 1984→2004(?). *Journal of Non-Crystalline Solids*, **73**, 599–612.

Dislich, H. (1971). New routes to multicomponent oxide glasses. *Angewandte Chemie, International Edition*, **10**, 363–70.

Dislich, H. & Hinz, P. (1982). History and principles of the sol–gel process and some new multicomponent oxide coatings. *Journal of Non-Crystalline Solids*, **48**, 11–16.

Dosch, R. G. (1984). Preparation of barium titanate films using sol–gel techniques. *Materials Research Society Symposium Proceedings*, **32**, 157–61.

Ebelman, J. J. (1846). Untersuchungen über die verbindungen der borsäure und kieselsäure mit aether. *Ann*, **57**, 319–55.

Ebelman, J. J. & Bouquet, M. (1846). Sur de nouvelles combinaisons de l'acide borique avec les éthers et sur l'éther sulfureux. *Ann. Chim. Phys.*, **17**, 54–73.

Ethyl Corporation. (1955). Preparation of alkali metal alcoholates. *United Kingdom Patent* 727,923.

Fegley, B. & Barringer, E. A. (1984). Synthesis, characterisation and processing of monosized ceramic powders. *Materials Research Society Symposium Proceedings*, **32**, 187–97.

Fegley, B., White, P. & Bowen, H. K. (1985). Processing and characterisation of ZrO_2 and Y-doped ZrO_2 powders. *American Ceramic Society Bulletin*, **64**, 1115–20.

Geffcken, W. & Berger, E. (1939). Verfahren zur änderung des reflexionsvermögens optischer gläser. *German Patent* 736,411.

Geotti-Bianchini, F., Guglielmi, M., Polato, P. & Soraru, G. D. (1984). Preparation and characterisation of Fe, Cr and Co oxide films on flat glass from gels. *Journal of Non-Crystalline Solids*, **63**, 251–9.

Gossink, R. G., Coenen, H. A. M., Engelfriet, A. R. C., Verheijke, M. L. & Verplane, J. C. (1975). Ultrapure SiO_2 and Al_2O_3 for the preparation of low-loss compounds glass. *Materials Research Bulletin*, **10**, 35–40.

Guglielmi, M. & Maddelena, A. (1985). Coating of glass fibres for cement composites by the sol–gel method. *Journal of Materials Science Letters*, **4**, 123–4.

Haaland, D. M. & Brinker, C. J. (1984). In-situ FT–IR studies of oxide and oxynitride sol–gel derived thin films. *Materials Research Society Symposium Proceedings* , **32**, 267–73.

References

Heistand II, R. H. & Chia, Y. H. (1986). Synthesis of submicron, narrow size distribution spherical zincite. *Materials Research Society Symposium Proceedings*, **73**, 93–8.

Hench, L. L. (1986). Use of drying control chemical additives (DCCAs) in controlling sol–gel processing. In *Science of Ceramic Chemical Processing*, ed. L. L. Hench & D. R. Ulrich, pp. 52–64. New York: John Wiley & Sons.

Hoch, M. & Nair, K. M. (1979). Preparation and characterisation of ultrafine powders of refractory nitrides. *American Ceramic Society Bulletin*, **58**, 187–90.

Huang, H. H., Orier, B. & Wilkes, G. L. (1985). Ceramers: Hybrid materials incorporating polymeric/oligomeric species with inorganic glasses by a sol–gel process. 2. Effect of acid content on the final properties. *Polymer Bulletin*, **14**, 557–64.

Iler, R. K. (1979). *The Chemistry of Silica*. New York: John Wiley & Sons.

Jean, J. H. & Ring, T. A. (1986). Nucleation and growth of monodispersed TiO_2 powders from alcohol solution in the presence of a sterically stabilising surfactant. *Proceedings of the British Ceramic Society*, **38**, 11–33.

Kamiya, K., Ohya, M. & Yoko, T. (1986). Nitrogen-containing SiO_2 glass fibres prepared by ammonolysis of gels made from silicon alkoxides. *Journal of Non-Crystalline Solids*, **83**, 208–22.

Kamiya, K., Yoko, T. & Bessho, M. (1987). Nitridation of TiO_2 fibres prepared by the sol–gel process. *Journal of Materials Science*, **22**, 937–41.

Keefer, K. D. (1984). The effect of hydrolysis conditions on the structure and growth of silicate polymers. *Materials Research Society Symposium Proceedings*, **32**, 15–24.

Kern, W. & Tracy, E. (1980). TiO_2 antireflection coating for silicon solar cells by spray deposition. *RCA Review*, **41**, 133–80.

Kistler, S. S. (1932). Coherent expanded aerogels. *Journal of Physical Chemistry*, **36**, 52–64.

Klein, L. C. (1984). Oxide coatings from the sol–gel process. *Ceramic Engineering and Science Proceedings*, **5**, 379–84.

Klein, L. C. (1985). Sol–gel processing of silicates. *Annual Review of Materials Science*, **15**, 227–48.

Klemperer, W. G., Mainz, V. V. & Millar, D. M. (1986). A solid state multinuclear magnetic resonance study of the sol–gel process using polysilicate precursors. *Materials Research Society Symposium Proceedings*, **73**, 15–25.

Komarneni, S. & Roy, R. (1985). Titania gel spheres by a new sol–gel process. *Materials Letters*, **3**, 165–7.

Kordas, G., Weeks, R. A. & Arfsten, N. (1985). Magnetic thin films produced by sol–gel processes. *Journal of Applied Physics*, **57**, 3812–3.

LaMer, V. K. & Dinegar, R. H. (1950). Theory, production and formation of monodispersed hydrosols. *Journal of the American Chemical Society*, **72**, 4847–54.

Lannutti, J. J. & Clark, D. E. (1984). Sol–gel derived coatings on SiC and silicate fibres. *Ceramic Engineering and Science Proceedings*, **5**, 574–82.

van Lierop, J. G., Huizing, A., Meerman, W. C. P. M. & Mulder, C. A. M. (1986). Preparation of dried monolithic SiO_2 gel bodies by an autoclave process. *Journal of Non-Crystalline Solids*, **82**, 265–70.

166

References

Liu, A. T. & Kleinschmit, P. (1986). Production of fumed oxides by flame hydrolysis. *Proceedings of the British Ceramic Society*, **38**, 1–10.

Loehman, R. E. (1985). Oxynitride glasses. In *Treatise on Materials Science and Technology*, vol. 26, ed. M. Tomozawa & R. H. Doremus, pp. 119–49. New York: Academic Press.

Makishima, A., Kubo, H., Wada, K., Kitami, Y. & Shimohira, T. (1986). Yellow coatings produced on glasses and aluminium by the sol–gel process. *Journal of the American Ceramic Society*, **69**, C127–9.

Martinsen, J., Figat, R. A. & Shafer, M. W. (1984). Preparation of thin composite coatings by sol–gel techniques. *Materials Research Society Symposium Proceedings*, **32**, 145–50.

Mazdiyasni, K. S. (1982). Powder synthesis from metal–organic precursors. *Ceramics International*, **8**, 42–56.

Mazdiyasni, K. S., Lynch, C. T. & Smith II, J. S. (1966). The preparation and some properties of yttrium, dysprosium and ytterbium alkoxides. *Inorganic Chemistry*, **5**, 342–6.

Mehrotra, R. C. (1983). Transition metal alkoxides. In *Advances in Inorganic Chemistry and Radiochemistry*, vol. 26, ed. J. H. Emeléus & A. G. Sharpe, pp. 269–335. London: Academic Press.

Mehrotra, R. C. (1954). The reaction of the alkoxides of titanium, zirconium and hafnium with esters. *Journal of the American Chemical Society*, **76**, 2266–7.

Mehrotra, R. C. (1953). Aluminium alkoxides. *Journal of the Indian Chemical Society*, **30**, 585–91.

Mehrotra, R. C. & Bhatnagar, D. D. (1965). Derivatives of antimony (III) with glycols. *Journal of the Indian Chemical Society*, **42**, 327–32.

Mehrotra, R. C. & Mittal, R. K. (1964). Synthesis of trialkylorthovanadates by alcohol interchange technique. *Zeitschrift für Anorganische und Allgemeine Chemie*, **327**, 311–4.

Milne, S. J. (1986). Preparation and ordering of spherical submicron monosized powders. *Proceedings of the British Ceramic Society*, **38**, 81–90.

Mitomo, M. & Yoshioka, Y. (1987). Process for the production of fine non-oxide powders from alkoxides. *United States Patent* 4,643,859.

Mukherjee, S. P. (1981). Gel-derived single-layer antireflection films with a refractive index gradient. *Thin Solid Films*, **81**, L89–90.

Mukherjee, S. P. & Lowdermilk, W. H. (1982). Gradient-index AR film deposited by the sol–gel process. *Applied Optics*, **21**, 293–6.

Mulder, C. A. M., van Leeuwen-Stienstra, G., van Lierop, J. G. & Woerdman, J. P. (1986). Chain-like structure of ultra-low density SiO_2 sol–gel glass observed by TEM. *Journal of Non-Crystalline Solids*, **82**, 148–53.

Ogihara, T., Mizutani, N. & Kato, M. (1987). Processing of monodispersed zirconia powders. *Ceramics International*, **13**, 35–40.

Okamura, H. & Bowen, H. K. (1986). Preparation of alkoxides for the synthesis of ceramics. *Ceramics International*, **12**, 161–71.

Orcel, G. & Hench, L. L. (1984). Use of a drying control chemical additive in the sol–gel processing of soda silicate and soda borosilicates. *Ceramic Engineering and Science Proceedings*, **5**, 546–55.

167

References

Perthuis, H. & Colomban, Ph. (1986). Sol–gel routes leading to Nasicon ceramics. *Ceramics International*, **12**, 39–52.

Philipp, G. & Schmidt, H. (1984). New materials for contact lenses prepared from Si- and Ti-alkoxides by the sol–gel process. *Journal of Non-Crystalline Solids*, **63**, 283–92.

Pope, J. M. & Harrison, D. E. (1981). Advanced method for making vitreous waste forms. In *Proceedings of the Third International Symposium on the Scientific Basis for Nuclear Waste Management*, vol. 3, ed. J. G. Moore, pp. 93–100. New York: Plenum Press.

Pouxviel, J. C., Boilot, J. P., Beloeil, J. C. & Lallemand, J. Y. (1987). NMR study of the sol–gel polymerization. *Journal of Non-Crystalline Solids*, **89**, 345–60.

Puyane, R. & Kato, I. (1983). Tin oxide films on glass substrates by a sol–gel technique. *Proceedings of SPIE–International Society for Optical Engineering*, **401**, 190–7.

Roy, D. M. & Roy, R. (1955). Synthesis and stability of minerals in the system $MgO-Al_2O_3-SiO_2-H_2O$. *American Mineralogist*, **40**, 147–78.

Roy, R. (1956). Aids in hydrothermal experimentation. II. Methods of making mixtures for both dry and wet phase equilibrium studies. *Journal of the American Ceramic Society*, **39**, 145–6.

Sakka, S. (1985). Sol–gel synthesis of glasses. Present and future. *American Ceramic Society Bulletin*, **64**, 1463–6.

Sakka, S. & Kamiya, K. (1982). The sol–gel transition in the hydrolysis of metal alkoxides in relation to the formation of glass fibres and films. *Journal of Non-Crystalline Solids*, **48**, 31–46.

Sakka, S., Kamiya, K., Makita, K. & Yamamoto, Y. (1983). Optical absorption of transition metal ions in silica coating films prepared by the sol–gel technique. *Journal of Materials Science Letters*, **2**, 395–6.

Scherer, G. W. (1987). Drying gels. IV. Cylinder and spheres. *Journal of Non-Crystalline Solids*, **91**, 101–21.

Schlichting, J. & Neumann, S. (1982). GeO_2-SiO_2 glasses from gels to increase the oxidation resistance of porous silicon containing ceramics. *Journal of Non-Crystalline Solids*, **48**, 185–94.

Schmidt, H. & Phillipp, G. (1985). Inorganic/organic polymers for contact lenses by the sol–gel process. *NATO ASI Ser Ser E*, **92**, 580–91.

Scholze, H. (1985). New possibilities for variation of glass structure. *Journal of Non-Crystalline Solids*, **73**, 669–80.

Schroeder, H. (1962). Properties and applications of oxide layers deposited on glass from organic solutions. *Optica Acta*, **9**, 249–54.

Schroeder, H. (1969). Oxide layer deposited from organic solutions. In *Physics of Thin Films*, vol. 3, ed. G. Hass & R. E. Thun, pp. 87–141. New York: Academic Press.

Shoup, R. D. & Wein, W. J. (1977). Low temperature production of high purity fused silica. *United States Patent* 4,059,658.

Silverman, L. A., Teowee, G. & Uhlmann, D. R. (1986). Characterisation of sol–gel derived tantalum oxide films. *Materials Research Society Symposium Proceedings*, **73**, 725–30.

References

Smith II, J. S., Dolloff, R. T. & Mazdiyasni, K. S. (1970). Preparation and characterisation of alkoxy-derived $SrZrO_3$ and $SrTiO_3$. *Journal of the American Ceramic Society*, **53**, 91–4.

Stöber, W., Fink, A. & Bohn, E. (1968). Controlled growth of monodispersed silica spheres in the micron size range. *Journal of Colloid and Interface Science*, **26**, 62–9.

Strawbridge, I. (1984). *Characterisation of Glassy, Thin Films and Bulk Materials Prepared by the Sol–Gel Process*, Ph.D. Thesis, University of Sheffield, England.

Strawbridge, I., Phalippou, J. & James, P. F. (1984). Characterisation of alkali alumino-borosilicate glass films prepared by the sol–gel process on window glass substrates. *Physics and Chemistry of Glasses*, **25**, 134–41.

Teichner, S. J. (1986). Aerogels of inorganic oxides. In *Aerogels*, ed. J. Fricke, pp. 22–30. Berlin: Springer-Verlag.

Teichner, S. J. (1953). Sur la préparation et quelques propriétés des alcoolates d'aluminium. *Compt. Rend.*, **237**, 810–2.

Tewari, P. H., Hunt, A. J., Lieber, J. G. & Lofftus, K. (1986). Microstructural properties of transparent silica aerogels. In *Aerogels*, ed. J. Fricke, pp. 142–7. Berlin: Springer-Verlag.

Thomas, I. M. (1977). Method for manufacturing silicate glasses from alkoxides. *United States Patent* 4,028,085.

Thomas, I. M. (1974). Method for producing glass precursor compositions and glass compositions therefrom. *United States Patent* 3,799,754.

Tormey, E. S., Pober, R. L., Bowen, H. K. & Calvert, P. D. (1984). Tape casting–future developments. In *Advances in Ceramics*, vol. 9, ed. J. A. Manges, pp. 140–9. Columbus, Ohio: American Ceramic Society Press.

Vance, E. R. (1986). Sol–gel production of titanosilicate glass-ceramics for nuclear waste immobilization. *Journal of Materials Science*, **21**, 1413–6.

Wallace, S. & Hench, L. L. (1984). The processing and characterisation of DCCA modified gel-derived silica. *Materials Research Society Symposium Proceedings*, **32**, 47–52.

Woignier, T., Phalippou, J. & Zarzycki, J. (1984). Monolithic aerogels in the systems $SiO_2–B_2O_3$, $SiO_2–P_2O_5$, $SiO_2–B_2O_3–P_2O_5$. *Journal of Non-Crystalline Solids*, **63**, 117–30.

Yamane, M., Aso, S. & Sakaino, T. (1978). Preparation of a gel from metal alkoxide and its properties as a precursor of oxide glass. *Journal of Materials Science*, **13**, 865–70.

Yoldas, B. E. (1973). Hydrolysis of aluminium alkoxides and bayerite conversion. *Journal of Applied Chemistry and Biotechnology*, **23**, 803–9.

Yoldas, B. E. (1975a). A transparent porous alumina. *American Ceramic Society Bulletin*, **54**, 286–8.

Yoldas, B. E. (1975b). Alumina sol preparation from alkoxides. *American Ceramic Society Bulletin*, **54**, 289–90.

Yoldas, B. E. (1977). Preparation of glasses and ceramics from metal–organic compounds. *Journal of Materials Science*, **12**, 1203–8.

Yoldas, B. E. (1986). Hydrolytic polycondensation of $Si(OC_2H_5)_4$ and effect of reaction parameters. *Journal of Non-Crystalline Solids*, **83**, 375–90.

References

Yoldas, B. E. & O'Keeffe, T. W. (1979). Antireflective coatings applied from metal–organic derived liquid precursors. *Applied Optics*, **18**, 3133–8.

Zarzycki, J. (1984). Monolithic xero- and aerogels for gel–glass processes. In *Ultrastructure Processing of Ceramics, Glasses and Composites*, ed. L. L. Hench & D. R. Ulrich, pp. 27–42. New York: John Wiley & Sons.

Zelinski, B. J. J. & Uhlmann, D. R. (1984). Gel technology in ceramics. *Journal of Physical Chemistry of Solids*, **45**, 1069–90.

Chapter 6

Billy, M. (1959). Préparation et définition du nitrure de silicium. *Annules de Chemie*, **4**, 795–851.

Billy, M., Brossard, M., Desmaison, J., Giraud, D. & Goursat, P. (1975). Synthesis of Si and Ge nitrides and Si oxynitride by ammonolysis of chlorides – comment on 'synthesis, characterisation and consolidation of Si_3N_4 obtained from ammonolysis of $SiCl_4$'. *Journal of the American Ceramic Society*, **58**, 254–5.

Billy, M., Jarige, J. & Progeas, L. (1970). Contribution a l'étude du système chlorure d'ammonium-ammoniac. *Bulletin de la Société Chimique de France*, **8–9**, 2924–31.

Crosbie, G. M. (1986). Preparation of silicon nitride powders. *Ceramic Engineering and Science Proceedings*, **7**, 1144–9.

Crosbie, G. M. (1987). Silicon nitride powder synthesis. In *Proceedings of the 24th Automotive Technology Development Contractors' Coordination Meeting*, pp. 255–65. Pennsylvania: Society of Automotive Engineers.

Eaborn, C. A. (1960). *Organosilicon Compounds*. London: Butterworths.

Franz, G., Schönfelder, L. & Wickel, U. (1986). Synthesis of fine grained Si_3N_4 powders by thermal decomposition of silicon diimide. In *Proceedings of the Second International Symposium on Ceramic Materials and Components for Engines*, Lübeck-Travemünde FRG, 14–17th April 1986, ed. W. Bunk & H. Hausner, pp. 117–24. Germany: Deutsche Keramische Gesellschaft.

Glemser, O. & Naumann, P. (1959). Ueber den thermischen abbau von siliciumdiimid $Si(NH)_2$. *Zeitschrift fur Anorganische und Allgemeine Chemie*, **298**, 134–41.

Iwai, T., Kawahito, T. & Yamada, T. (1980). Process for producing metallic nitride powders. *United States Patent* 4,196,178.

Johnson, C. E., Hickey, D. K. & Harris, D. C. (1986). Synthesis of metal sulphide powders from organometallics. *Materials Research Society Symposium Proceedings*, **73**, 785–9.

Kalyoncu, R. S. (1985). BN powder synthesis at low temperatures. *Ceramic Engineering and Science Proceedings*, **6**, 1356–64.

Kohtoku, Y., Yamada, T., Miyazaki, H. & Iwai, T. (1986). The development of ceramics from amorphous silicon nitride. In *Proceedings of the Second International Symposium on Ceramic Materials and Components for Engines*, Lübeck-Travemünde FRG, 14–17th April 1986, ed. W. Bunk & H. Hausner, pp. 101–8. Germany: Deutsche Keramische Gesellschaft.

References

Larsson, E. & Bjellerup, L. (1953). The reaction of diphenyldichlorosilane with ammonia and amines. *Journal of the American Chemical Society*, **75**, 995–7.

Lengfeld, F. (1899). The action of ammonia and amines on chlorides of silicon. *American Chemical Journal*, **21**, 531–7.

Mazdiyasni, K. S. & Cooke, C. M. (1973). Synthesis, characterisation and consolidation of Si₃N₄ obtained from ammonolysis of SiCl₄. *Journal of the American Ceramic Society*, **56**, 628–33.

Mazdiyasni, K. S. & Cooke, C. M. (1976). Synthesis of high purity, alpha phase silicon nitride powder. *United States Patent* 3,959,446.

Mazdiyasni, K. W., West, R. & David, L. D. (1978). Characterisation of organosilicon-infiltrated porous reaction-sintered Si₃N₄. *Journal of the American Ceramic Society*, **61**, 504–8.

Melling, P. J. (1984). Alternative methods of preparing chalcogenide glasses. *American Ceramic Society Bulletin*, **63**, 1427–9.

Persoz, M. (1830). Observations sur les combinaisons du gaz ammoniac avec les chlorures metalliques. *Ann. Chim. Phys.*, **44**, 315–25.

Ritter, J. J. (1986). A low temperature chemical route to precursors of boride and carbide ceramic powders. *Materials Research Society Symposium Proceedings* , **73**, 367–72.

Segal, D. L. (1986). A review of preparative routes to silicon nitride powders. *Transactions of the Journal of the British Ceramic Society*, **85**, 184–7.

Seyferth, D., Wiseman, G. H. & Prud'homme, C. (1983). A liquid silazane precursor to silicon nitride. *Journal of the American Ceramic Society*, **66**, C13–4.

Seyferth, D., Wiseman, G.H. & Prud'homme, C. (1984). Silicon–nitrogen polymers and ceramics derived from reactions of dichlorosilane, H₂SiCl₂. *Materials Science Research*, **17**, 263–9.

Stock, A. & Somieski, K. (1921). Siliciumwasserstoffe, X. Stickstoffhaltige verbindungen. *Ber. Deutsch Chem. Ges.*, **54**, 740–58.

Vigoureux, E. & Hugot, C. R. (1903). Sur l'amidure et l'imidure de silicium. *Hebd. Séances Acad. Sci.*, **136**, 1670–72.

Yamada, T., Kawahito, T. & Iwai, T. (1984). Preparation of silicon nitride powder by imide decomposition method. In *Proceedings of the First International Symposium on Ceramic Components for Engines*, ed. S. Sōmiya, E. Kanai & K. Ando, pp. 333–42. Tokyo: KTK Scientific Publishers.

Chapter 7

Andersson, C. H. & Warren, R. (1984). Silicon carbide fibres and their potential for use in composite materials. *Composites*, **15**, 16–24.

Baney, R. H., Gaul, Jr, J. H. & Hilty, T. K. (1984). The conversion of methylchloropolysilanes and polydisilylazanes to silicon carbide and silicon carbide/silicon nitride ceramics respectively. *Materials Science Research*, **17**, 253–62.

Burkhard, C. A. (1949). Polydimethylsilanes. *Journal of The American Chemical Society*, **71**, 963–4.

References

Coblenz, W. S., Wiseman, G. H., Davies, P. B. & Rice, R. W. (1984). Formation of ceramic composites and coatings utilizing polymer pyrolysis. *Materials Science Research*, **17**, 271–85.

David, L. D. (1987). The discovery of soluble polysilane polymers. *Chemistry in Britain*, **23**, 553–6.

Fritz, G. & Matern, E. (1986). *Carbosilanes: Syntheses and Reactions*. Berlin: Springer-Verlag.

Ishikawa, T. & Teranishi, H. (1981). Research and development on silicon carbide continuous fibre Nicalon. *New Materials and New Processes*, **1**, 36–41.

Kipping, F. S. & Sands, J. E. (1921). Organic derivatives of silicon. Part XXV. Saturated and unsaturated silicohydrocarbons, Si_4Ph_8. *Journal of the Chemical Society*, **119**, 830–47.

Lee, J. G. & Cutler, I. B. (1975). Formation of silicon carbide from rice hulls. *American Ceramic Society Bulletin*, **54**, 195–8.

Legrow, G. E., Lim, T. F., Lipowitz, J. & Reaoch, R. S. (1987). Ceramics for hydridopolysilazane. *American Ceramic Society Bulletin*, **66**, 363–7.

Mazdiyasni, K. S., West, R. & David, L. D. (1978). Characterisation of organosilicon-infiltrated porous reaction-sintered Si_3N_4. *Journal of the American Ceramic Society*, **61**, 504–8.

Narula, C. K., Paine, R. T. & Schaeffer, R. (1986). Precursors to boron nitrogen macromolecules and ceramics. *Materials Research Society Symposium Proceedings*, **73**, 383–8.

Okamura, K. (1987). Ceramic fibres from polymer precursors. *Composites*, **18**, 107–20.

Okamura, K., Sato, M., Hasegawa, Y. & Amano, T. (1984). The synthesis of silicon oxynitride fibres by nitridation of polycarbosilane. *Chemistry Letters*, pp. 2059–60.

Penn, B. G., Ledbetter III, F. E., Clemons, J. M. & Daniels, J. G. (1982). Preparation of silicon carbide-silicon nitride fibres by the controlled pyrolysis of polycarbosilazane precursors. *Journal of Applied Polymer Science*, **27**, 3751–61.

Teranishi, H., Ichikawa, H. & Ishikawa, T. (1983). Properties of silicon carbide fibre 'Nicalon' and its aluminium composites. *New Materials and New Processes*, **2**, 379–85.

Walker, Jr, B. E., Rice, R. W., Becher, P. F., Bender, B. A. & Coblenz, W. S. (1983). Preparation and properties of monolithic and composite ceramics produced by polymer pyrolysis. *American Ceramic Society Bulletin*, **62**, 916–23.

West, R. C. (1981). Phenylmethylpolysilane polymers and process for their preparation. *United States Patent* 4,260,780.

West, R. (1984). Polysilane precursors to silicon carbide. In *Ultrastructure Processing of Ceramics, Glasses and Composites*, ed. L. L. Hench & D. R. Ulrich, pp. 235–44. New York: John Wiley & Sons.

West, R. (1986). The polysilane high polymers. *Journal of Organometallic Chemistry*, **300**, 327–46.

West, R. C., David, L. D., Djurovich, P. I., Yu, H. & Sinclair, R. (1983). Polysilastyrene: Phenylmethylsilane-dimethylsilane copolymer as precursors to silicon carbide. *American Ceramic Society Bulletin*, **62**, 899–903.

References

Yajima, S., Hasegawa, Y., Hayashi, J. & Imura, M. (1978b). Synthesis of continuous silicon carbide fibre with high tensile strength and high Young's modulus. Part I. Synthesis of polycarbosilane as precursor. *Journal of Materials Science*, **13**, 2569–76.

Yajima, S., Hayashi, J. & Okamura, K. (1977). Pyrolysis of a polyborodiphenylsiloxane. *Nature*, **266**, 521–2.

Yajima, S., Hayashi, J. & Omori, M. (1978a). Silicon carbide fibres having a high strength and a method for producing said fibres. *United States Patent* 4,100,233.

Yajima, S., Hayashi, J. & Omori, M. (1978c). Method for producing silicon carbide sintered moldings consisting mainly of SiC. *United States Patent*, 4,122,139.

Yajima, S., Hayashi, J. & Omori, M. (1975a). Continuous silicon carbide fibre of high tensile strength. *Chemistry Letters*, pp. 931–4.

Yajima, S., Hayashi, J., Omori, M. & Okamura, K. (1976b). Development of a silicon carbide fibre with high tensile strength. *Nature*, **261**, 683–5.

Yajima, S., Iwai, T., Yamamura, T., Okamura, K. & Hasegawa, Y. (1981). Synthesis of a polytitanocarbosilane and its conversion into inorganic compounds. *Journal of Materials Science*, **16**, 1349–55.

Yajima, S., Liaw, C., Omori, M. & Hayashi, J. (1976c). Molecular weight distribution of polycarbosilane as a starting material of the silicon carbide fibre with high tensile strength. *Chemistry Letters*, pp. 435–6.

Yajima, S., Okamura, K. & Hayashi, J. (1975b). Structural analysis in continuous silicon carbide fibre of high tensile strength. *Chemistry Letters*, pp. 1209–12.

Yajima, S., Okamura, K., Hayashi, J. & Omori, M. (1976d). Synthesis of continuous SiC fibres with high tensile strength. *Journal of the American Ceramic Society*, **59**, 324–7.

Yajima, S., Omori, M., Hayashi, J., Okamuya, K., Matsuzawa, T. & Liaw, C. (1976a). Simple synthesis of the continuous SiC fibre with high tensile strength. *Chemistry Letters*, pp. 551–4.

Yajima, S., Shishido, T. & Hamano, M. (1977). SiC and Si_3N_4 sintered bodies with new borodiphenylsiloxane polymers as binder. *Nature*, **266**, 522–4.

Chapter 8

Anonymous. (1986). Commercialization of hydrothermal technique for producing superfine zirconia. *Techno Japan*, **19**, 67–8.

Brace, R. & Matijević, E. (1973). Aluminium hydrous oxide sols. I. Spherical particles of narrow size distribution. *Journal of Inorganic and Nuclear Chemistry*, **35**, 3691–705.

Demchak, R. & Matijević, E. (1969). Preparation and particle size analysis of chromium hydroxide hydrosols of narrow size distributions. *Journal of Colloid and Interface Science*, **31**, 257–62.

Hill, C. G. A. (1986). Inorganic phosphors. In *Fine Chemicals for the Electronics Industry*, ed. P. Bamfield, pp. 159–75. Dorchester: Royal Society of Chemistry.

References

Hoffman, D. W., Roy, R. & Komarneni, S. (1984). Diphasic xerogels, a new class of materials: Phases in the system Al_2O_3–SiO_2. *Journal of the American Ceramic Society*, **67**, 468–71.

Komarneni, S., Roy, R., Breval, E., Ollinen, M. & Suwa, Y. (1986). Hydrothermal route to ultrafine powders utilising single and diphasic gels. *Advanced Ceramic Materials*, **1**, 87–92.

Kutty, T. R. N., Tareen, J. A. K., Basavalingu, B. & Puttaraju, B. (1982). Low temperature hydrothermal reduction of metal hydroxides to metal powders. *Materials Letters*, **1**, 67–70.

Kutty, T. R. N. & Vivekanandan, R. (1987). Preparation of $CaTiO_3$ fine powders by the hydrothermal method. *Materials Letters*, **5**, 79–83.

Matijević, E. (1981). Monodispersed metal (hydrous) oxides—A fascinating field of colloid science. *Accounts of Chemical Research*, **14**, 22–9.

Matijević, E. (1984). Monodispersed colloidal metal oxides, sulphides and phosphates. In *Ultrastructure Processing of Ceramics, Glasses and Composites*, ed. L. L. Hench & D. R. Ulrich, pp. 334–52. New York: John Wiley & Sons.

Matijević, E. (1987). Colloid science in ceramic powders preparation. In *Materials Science Monographs: High Tech Ceramics*, vol. 38, part A, ed. P. Vincenzini, pp. 441–58. Amsterdam: Elsevier.

Matijević, E., Budnik, M. & Meites, L. (1977). Preparation and mechanism of formation of titanium dioxide hydrosols of narrow size distribution. *Journal of Colloid and Interface Science*, **61**, 302–11.

Matijević, E., Simpson, C. M., Amin, N. & Arajs, S. (1986). Preparation and magnetic properties of well-defined colloidal chromium ferrites. *Colloids and Surfaces*, **21**, 101–8.

Monhemius, A. J. & Steele, B. C. H. (1986). Oxide powders by hydrothermal synthesis. *Proceedings of the British Ceramic Society*, **38**, 35.

Sōmiya, S. (1984a). Hydrothermal preparation and sintering of fine ceramic powders. *Materials Research Society Symposium Proceedings*, **24**, 255–71.

Sōmiya, S. (1984b). Hydrothermal reaction sintering of oxides. In *Sintered Metal Ceramic Composites*, ed. G. S. Upadhyaya, pp. 97–103. Amsterdam: Elsevier.

Stambaugh, E. P., Adair, J. H., Sekercioglu, I. & Wills, R. R. (1986). Hydrothermal method for producing stabilized zirconia. *United States Patent*, 4,619,817.

Takamori, T. & David, L. D. (1986). Controlled nucleation for hydrothermal growth of yttrium aluminium garnet powders. *American Ceramic Society Bulletin*, **65**, 1282–6.

Tani, E., Yoshimura, M. & Sōmiya, S. (1983). Formation of ultrafine tetragonal ZrO_2 powder under hydrothermal conditions. *Journal of the American Ceramic Society*, **66**, 11–14.

Toraya, H., Yoshimura, M. & Sōmiya, S. (1983). Hydrothermal oxidation of Hf metal chips in the preparation of monoclinic HfO_2 powders. *Journal of the American Ceramic Society*, **66**, 148–50.

References

Vivekanandan, R., Philip, S. & Kutty, T. R. N. (1986). Hydrothermal preparation of Ba(Ti,Zr)O$_3$ fine powders. *Materials Research Bulletin*, **22**, 99–108.

Yoshimura, M. & Sōmiya, S. (1982). Synthesis and sintering of zirconia fine powders by hydrothermal reactions from zirconium metal and high-temperature high-pressure solutions. In *Materials Science Monographs: Sintering, Theory and Practice*, vol. 14, ed. D. Kolar, S. Pejovnik & M. M. Ristic, pp. 417–22. Amsterdam: Elsevier.

Chapter 9

Cannon, W. R., Danforth, S. C., Flint, J. H., Haggerty, J. S. & Marra, R. A. (1982a). Sinterable ceramic powders from laser-driven reactions. I. Process description and modelling. *Journal of the American Ceramic Society*, **65**, 324–30.

Cannon, W. R., Danforth, S. C., Haggerty, J. S. & Marra, R. A. (1982b). Sinterable ceramic powders from laser-driven reactions. II. Powder characteristics and process variables. *Journal of the American Ceramic Society*, **65**, 330–5.

Canteloup, J. & Mocellin, A. (1975). Synthesis of ultrafine nitrides and oxynitrides in an rf plasma. *Proceedings of the British Ceramic Society*, **6**, 209–21.

Carbone, T. J. & Rossing, B. R. (1986). SiC ceramics produced from plasma synthesized powders. In *Proceedings of the Second International Symposium on Ceramic Materials and Components for Engines*, Lübeck-Travemünde FRG, 14–17th April 1986, ed. W. Bunk & H. Hausner, pp. 47–54. Germany: Deutsche Keramische Gesellschaft.

Danforth, S. C. & Haggerty, J. S. (1983). Mechanical properties of sintered and nitrided laser-synthesized silicon powder. *Journal of the American Ceramic Society*, **66**, C-58–9.

Defay, R., Prigogine, I., Bellemans, A. & Everett, D. H. (1966). *Surface Tension and Adsorption*. London: Longman.

Economy, J., Smith, W. D. & Lin, R. Y. (1973). Preparation of refractory fibres by chemical conversion method. *Applied Polymer Symposium*, **21**, 131–41.

Fauchais, P., Boudrin, E., Coudert, J. F. & McPherson, R. (1983). High pressure plasmas and their application to ceramic technology. In *Topics in Current Chemistry*, vol. 107, ed. S. Veprek & M. Venugoplan, pp. 59–183. German Democratic Republic: Springer Verlag.

Gambling, W. A. (1986). Glass, light and the information revolution. *Glass Technology*, **27**, 179–87.

Gani, M. S. J. & McPherson, R. (1987). Al$_2$O$_3$–SnO$_2$ composite plasma-synthesized powders. *Journal of Materials Science Letters*, **6**, 681–2.

Gani, M. S. J. & McPherson, R. (1974). Structure of Al$_2$O$_3$–TiO$_2$ sub-micron particles. In *8th International Congress on Electron Microscopy*, 25–31st August 1974, vol. 1, ed. J. V. Sanders & D. J. Goodchild, pp. 534–5. Canberra: Australian Academy of Science.

References

Goodwin, C. A. (1982). The use of silicon nitride in semiconductor devices. *Ceramic Engineering & Science Proceedings*, **3**, 109–19.

Haggerty, J. S., Flint, J. H., Garvey, G. J., Lihrmann, J. M. & Ritter, J. E. (1986). High strength oxidation resistant reaction-bonded silicon nitride from laser synthesized silicon powder. In *Proceedings of the Second International Symposium on Ceramic Materials and Components for Engines*, Lübeck-Travemünde FRG, 14–17th April 1986, ed. W. Bunk & H. Hausner, pp. 147–54. Germany: Deutsche Keramische Gesellschaft.

Kato, A., Hojo, J. & Okabe, Y. (1981). Formation of ultrafine powders of refractory nitrides and carbides by vapour phase reaction. *Memoirs of the Faculty of Engineering Kyushu University*, **41**, 319–34.

Kato, A., Hojo, J. & Watari, T. (1984). Some common aspects of the formation of non-oxide powders by the vapour reaction method. *Materials Science Research*, **17**, 123–35.

Kennedy, P. & North, B. (1983). The production of fine silicon carbide powder by the reaction of gaseous silicon monoxide with particulate carbon. *Proceedings of the British Ceramic Society*. **33**, 1–15.

Kizaki, Y., Kandori, T. & Fujitani, Y. (1985). Synthesis and characterisation of Si_3N_4 powder produced by laser-induced chemical reactions. *Japanese Journal of Applied Physics*, **24**, 800–5.

Komeya, K. & Inoue, H. (1975). Synthesis of the α form of silicon nitride from silica. *Journal of Materials Science*, **10**, 1243–6.

Komeya, K., Inoue, H., Matake, S. & Endo, H. (1984). Method of making α-silicon nitride. *United States Patent* 4,428,916.

Kong, P. C. & Pfender, E. (1987). Formation of ultrafine β-silicon carbide powders in an argon thermal plasma jet. *Langmuir*, **3**, 259–65.

Lumby, R. J. (1976). Manufacture of silicon nitride powder. *United States Patent* 3,937,792.

McPherson, R. (1973). The structure of Al_2O_3–Cr_2O_3 powders condensed from a plasma. *Journal of Materials Science*, **8**, 859–62.

McPherson, R. & Aik, K. K. (1987). Plasma synthesis of zirconia-based composite powders. *Paper presented at the 8th International Symposium on Plasma Chemistry*, Tokyo, September 1987, in press.

Morgan, P. E. D. (1980). The α/β-Si_3N_4 question. *Journal of Materials Science*, **15**, 791–3.

Morgan, P. E. D. (1983). Studies for the production of super-pure silicon nitride powder. *United States Department of Energy Report* DE 83-004383.

Morgan, P. E. D. & Pugar, E. A. (1985a). Synthesis of Si_3N_4 with emphasis of Si–S–N chemistry. *Journal of the American Ceramic Society*, **68**, 699–703.

Morgan, P. E. D. & Pugar, E. A. (1985b). Process for producing amorphous and crystalline silicon nitride. *United States Patent* 4,552,740.

Prochaska, S. & Greskovich, C. (1978). Synthesis and characterisation of a pure silicon nitride powder. *American Ceramic Society Bulletin*, **57**, 579–81.

Ramsay, J. D. F. & Avery, R. G. (1977). Improvements in or relating to the preparation of carbide powders. *United Kingdom Patent* 1,479,727.

Rice, G. W. (1986). Laser synthesis of Si/C/N powder from 1,1,1,3,3,3-hexamethyldisilazane. *Journal of the American Ceramic Society*, **69**, C-183–5.

References

Schwier, G. (1983). The preparation of fine silicon nitride powders. In *Progress in Nitrogen Ceramics*, ed. F. L. Riley, pp. 157–66. The Hague: Martinus Nijhoff.

Segal, D. L. (1986). A review of preparative routes to silicon nitride powders. *Transactions of the Journal of the British Ceramic Society*, **85**, 184–7.

Sheppard, L. M. (1987). Vapour-phase synthesis of ceramics. *Advanced Materials and Processes Incorporating Metal Progress*, April 1987, pp. 53–8.

Strickland-Constable, R. F. (1968). *Kinetics and Mechanism of Crystallization*. London: Academic Press.

Symons, W., Nilsen, K. J. & Danforth, S. C. (1986). Synthesis and processing of laser synthesized silicon nitride powders. In *Proceedings of the Second International Symposium on Ceramic Materials and Components for Engines*. Lübeck-Travemünde FRG, 14–17 April 1986, ed. W. Bunk & H. Hausner, pp. 39–46. Germany: Deutsche Keramische Gesellschaft.

Ulrich, G. D. (1984). Flame synthesis of fine particles. *Chemical and Engineering News*, August 1984, pp. 22–9.

Vogt, G. J., Hollabaugh, C. M., Hull, D. E., Newkirk, L. R. & Petrovic, J. J. (1984). Novel rf plasma system for the synthesis of ultrafine ultrapure SiC and Si₃N₄. *United States Department of Energy Report DE 84-003774*. Los Alamos: Department of Energy.

Vogt, G. J., Vigil, R. S., Newkirk, L. R. & Trkula, M. (1985). Synthesis of ultrafine ceramic and metallic powders in a thermal argon rf plasma. In *Proceedings of the 7th International Symposium on Plasma Chemistry*, Eindhoven, 1–5 July 1985, vol. 2, ed. C. J. Timmermans, pp. 668–73. Eindhoven: International Union of Pure and Applied Chemistry.

Zhu, C. W. & Yan, J. P. (1985). Preparation of ultrafine Si₃N₄ powders in radio-frequency plasma. In *Proceedings of the 7th International Symposium on Plasma Chemistry*, Eindhoven, 1–5 July 1985, vol. 2, ed. C. J. Timmermans, pp. 657–61. Eindhoven: International Union of Pure and Applied Chemistry.

Chapter 10

Anderton, D. J. & Sale, F. R. (1979). Production of strontium-doped lanthanum cobaltite conducting oxide powder by freeze-drying technique. *Powder Metallurgy*, **1**, 8–13.

Barboux, P., Tarascon, J. M., Bagley, B. G., Greene, L. H. & Hull, G. W. (1987). The preparation of bulk and thick films of YBa₂Cu₃O₇₋δ using solution techniques. Paper number AA2.3 presented at a *Symposium on High-Temperature Superconductors*, Boston, Massachusetts, 30 November–5 December 1987. To be published in The Materials Research Society Symposium Proceedings, vol. 99.

Dole, S. L., Scheidecker, R. W., Shiers, L. E., Berard, M. F. & Hunter, Jr, O. (1978). Techniques for preparing highly sinterable oxide powders. *Materials Science and Engineering*, **32**, 277–81.

References

van de Graaf, M. A. C. G. & Burggraaf, A. J. (1984). Wet-chemical preparation of zirconia powders. Their microstructure and behaviour. In *Advances in Ceramics*, vol. 12, ed. N. Clausen, M. Rühle & A. H. Heuer, pp. 744–65. Columbus, Ohio: The American Ceramic Society.

Li, L. E., Ni, H. Y. & Yin, Z. E. (1983). Preparation of active powders for electronic ceramics by alcohol dehydration of citrate solutions. In *Materials Science Monographs: Ceramic Powders, Preparation, Consolidation and Sintering*, vol. 16, ed. P. Vincenzini, pp. 593–600. Amsterdam: Elsevier.

Lucchini, E., Meriani, S., Delben, F. & Paoletti, S. (1984). A new method for low temperature preparation of barium hexaferrite powders. *Journal of Materials Science*, **19**, 121–4.

Mahloojchi, F., Sale, F. R., Ross, J. W. & Shah, N. J. (1987). Oxide superconductors produced by the citrate gel process. Paper number P18 presented at *The First European Workshop on High T_c Superconductors and Potential Applications*, Genoa, Italy, 1–3 July, 1987.

Marcilly, C., Courty, P. & Delmon, B. (1970). Preparation of highly dispersed mixed oxides and oxide solid solutions by pyrolysis of amorphous organic precursors. *Journal of The American Ceramic Society*, **53**, 56–7.

Matson, S. W., Petersen, R. C. & Smith, R. D. (1986a). Formation of silica powders from the rapid expansion of supercritical solutions. *Advanced Ceramic Materials*, **1**, 242–6.

Matson, D. W., Petersen, R. C. & Smith, R. D. (1986b). The preparation of polycarbosilane powders and fibres during rapid expansion of supercritical fluid solutions. *Materials Letters*, **4**, 429–32.

Mazdiyasni, K. S. (1982). Powder synthesis from metal–organic precursors. *Ceramics International*, **8**, 42–56.

Mulder, B. J. (1970). Preparation of $BaTiO_3$ and other ceramic powders by co-precipitation of citrates in an alcohol. *American Ceramic Society Bulletin*, **49**, 990–3.

Nielsen, F. (1982). Spray drying pharmaceuticals. *Manufacturing Chemist*, **53**, 38–41.

Real, M. W. (1986). Freeze-drying alumina powders. *Proceedings of The British Ceramic Society*, **38**, 59–66.

Tang, Y., Ling, B., Zhang, W., Liu, Z., Zheng, X., Wu, N., Shao, M., Li, C., Chen, K. & Li, J. (1987). Preparative and structural studies on the superconducting phase $YBa_2Cu_3O_{7-\delta}$. In *Proceedings of the Beijing International Workshop on High Temperature Superconductivity*, Beijing. People's Republic of China, 29 June–1 July 1987, ed. Z. Z. Gan, G. J. Cui, G. Z. Yang & Q. S. Yang, pp. 129–44. Singapore: World Scientific Publishing Company Limited.

Appendix

Allen, T. (1981). *Particle Size Measurement*, 3rd edition. London: Chapman and Hall.

Bacon, G. E. (1977). *Neutron Scattering in Chemistry*. London: Butterworths.

178

References

Berne, B. J. & Pecora, R. (1976). *Dynamic Light Scattering with Applications to Chemistry, Biology and Physics*. New York: John Wiley & Sons, Inc.

Bowen, T. J. (1970). *An Introduction to Ultracentrifugation*. London: John Wiley & Sons.

Bristow, P. A. (1976). *Liquid Chromatography in Practice*. Cheshire: HETP.

Brunauer, S., Emmett, P. H. & Teller, E. (1938). Adsorption of gases in multimolecular layers. *Journal of the American Chemical Society*, **60**, 309–19.

Chu, B. (1974). *Laser Light Scattering*. New York: Academic Press.

Fryer, J. R. (1979). *The Chemical Applications of Transmission Electron Microscopy*. London: Academic Press.

Gregg, S. J. (1986). Sixty years in the physical adsorption of gases. *Colloids and Surfaces*, **21**, 109–24.

Guinier, A. & Fournet, G. (1955). *Small Angle Scattering of X-Rays*. New York: John Wiley & Sons.

Kerker, M. (1969). *The Scattering of Light and Other Electromagnetic Radiation*. New York: Academic Press.

Klug, H. P. & Alexander, L. E. (1974). *X-ray Diffraction Procedure for Polycrystalline and Amorphous Materials*, 2nd edition. New York: John Wiley & Sons.

Knox, J. H. (1976). Theory of HPLC. Part I. With special reference to column design. In *Practical High Performance Liquid Chromatography*, ed. C. F. Simpson, pp. 19–46. London: Heyden & Son Limited.

Lloyd, P. J. & Treasure, C. R. G. (1976). Particle size analysis. In *International Review of Science, Physical Chemistry Series Two. Analytical Chemistry–Part I*, vol. 12, ed. T. S. West, pp. 289–325. London: Butterworths.

Ottewill, R. H. (1982). Small angle neutron scattering. In *Colloidal Dispersions*, ed. J. W. Goodwin, pp. 143–63. London: Royal Society of Chemistry.

Porod, G. (1982). General theory. In *Small Angle X-Ray Scattering*, ed. O. Glatter & O. Kratky, pp. 17–51. London: Academic Press.

Pusey, P. N. (1982). Light scattering. In *Colloidal Dispersions*, ed. J. W. Goodwin, pp. 129–42. London: Royal Society of Chemistry.

Reimer, L. (1985). *Scanning Electron Microscopy. Physics of Image Formation and Microanalysis*. Berlin: Springer Verlag.

Rideal, G. (1982). Making a start in particle size analysis. *Laboratory Equipment Digest*, June 1982, pp. 83–7.

Sing, K. S. W., Everett, D. H., Haul, R. A. W., Moscou, L., Pierotti, R. A., Rouquerol, J. & Siemieniewska, T. (1985). Reporting physisorption data for gas/solid systems with special reference to the determination of surface area and porosity. *Pure and Applied Chemistry*, **57**, 603–19.

Society for Analytical Chemistry. (1968). *The Determination of Particle Size. I. A Critical Review of Sedimentation Methods*. London: Society for Analytical Chemistry.

Veale, C. R. (1972). *Fine Powders, Preparation, Properties and Uses*. London: Applied Science Publishers.

Weiner, B. B. (1984). Particle sizing using photon correlation spectroscopy. In *Modern Methods of Particle Size Analysis*, ed. H. G. Barth, pp. 93–116. New York: John Wiley & Sons.

Index

Index

Index

Printed in the United States
By Bookmasters